A MASTER GUIDE TO GRMS RADIO

Comprehensive Techniques and Strategies for Ensuring Safe and Effective Communication for Beginners and Pros

I0477590

WisdomBytes Solutions

TABLE OF CONTENTS

CHAPTER 1

INTRODUCTION TO GMRS RADIOS

Welcome to the dynamic world of GMRS radios! This chapter serves as your comprehensive guide, exploring everything you need to know about this versatile two-way communication technology. Whether you're a seasoned outdoors enthusiast, a business professional seeking reliable on-site coordination, or simply someone intrigued by the possibilities of off-grid communication, GMRS radios offer a unique solution.

As we dig into and explore this medium, we'll begin by unpacking the fundamentals of GMRS in Section 1.1, exploring its technical definition and core functionalities. Section 1.2 will then take a historical detour, tracing the evolution of GMRS and its continuous advancements. Understanding its origins sheds light on the ongoing development that keeps GMRS relevant in today's ever-changing technological landscape.

Next, Section 1.3 dives into the heart of the matter: the importance and diverse applications of GMRS radios. We'll explore how this technology empowers individuals and groups in various scenarios, from enhancing safety during outdoor adventures to streamlining communication within businesses.

As with any radio service, regulations play a crucial role. Section 1.4 will equip you with the knowledge you need regarding GMRS licensing requirements. We'll navigate the legalities, ensuring you operate your GMRS radio within the proper framework. Finally, Section 1.5 looks into the technical side of the spectrum, explaining how GMRS utilizes specific frequencies and how these allocations work to prevent interference.

By the end of this chapter, you'll possess a thorough understanding of GMRS radios, their historical significance, their practical applications, and the regulatory landscape surrounding their use. So, prepare to unlock a new dimension of communication – let's explore the exciting world of GMRS radios!

1.1 What is GMRS?

At its core, GMRS (General Mobile Radio Service) is a licensed two-way radio communication system designed for short-range voice interactions. Operating on designated Ultra High Frequency (UHF) bands, GMRS radios provide a reliable and independent communication channel, particularly useful in situations where traditional cell phone service might be limited or unavailable.

Unlike citizen band (CB) radios or walkie-talkies, GMRS radios necessitate a Federal Communications Commission (FCC) license for operation in the United States. This licensing process ensures responsible use of the allocated frequencies and minimizes interference within the GMRS spectrum.

Technically, GMRS radios function by transmitting and receiving radio waves within a specific range of UHF frequencies, typically between 462 MHz and 467 MHz. These radio waves travel in a relatively straight line, meaning communication range is generally limited by factors like terrain and distance. In ideal conditions, GMRS radios can offer a range of several miles, although obstacles and environmental conditions can affect this.

Here's a breakdown of some key characteristics of GMRS radios:

- **Two-way Communication:** Unlike one-way paging systems, GMRS radios facilitate interactive conversations between two or more radio units tuned to the same channel.
- **Push-to-Talk (PTT) Operation:** Similar to walkie-talkies, GMRS radios require users to press a button to initiate transmission, preventing accidental broadcasts and allowing for clear channel coordination.
- **Limited Data Capabilities:** While primarily designed for voice communication, some GMRS radios offer basic data transmission features like text messaging or rudimentary GPS location sharing.
- **Analog vs. Digital Technology:** The majority of GMRS radios utilize analog technology, offering a familiar and reliable method of communication. However, digital GMRS radios are becoming increasingly available, potentially providing clearer audio and enhanced features in the future.

Understanding these core functionalities gives you a solid foundation for appreciating the potential of GMRS radios. The following sections will explore the historical context of GMRS, its diverse applications in various fields, and the regulatory framework surrounding its use.

1.2 History and Evolution of GMRS: A Chronicle of Innovation and Communication Needs

The story of GMRS radios stretches back further than a simple origin point in the 1960s. It's a narrative woven from the pioneering spirit of early radio enthusiasts, the ever-present need for reliable communication, and the continuous march of technological advancements. To fully appreciate the capabilities of GMRS today, we must look into this rich history, exploring the stepping stones that led to the robust system we know now.

Seeds of Innovation: The Dawn of Two-Way Radio (1920s-1950s)

While the official birth of GMRS can be traced to the 1960s, its roots can be found much earlier, nestled within the broader development of two-way radio technology. The early

20th century saw a surge in experimentation with radio waves, with pioneers like Guglielmo Marconi and Reginald Fessenden laying the groundwork for practical applications. By the 1920s, rudimentary two-way radio systems emerged, primarily used for communication between ships and shore stations or for early police dispatch systems. These early systems were bulky, complex, and often unreliable, but they laid the foundation for future developments.

The Rise of Citizen Band Radio: A Catalyst for Change (1940s-1960s)

The concept of citizen band (CB) radio emerged in the 1940s, fueled by a desire for readily accessible two-way communication for the general public. Initially envisioned for emergency preparedness and disaster relief, CB radios quickly gained popularity among hobbyists and enthusiasts. However, the limited number of channels and the lack of licensing requirements led to congestion and interference, hindering the effectiveness of the system.

Introducing Class A Citizens Radio Service: A Flawed, Yet Pioneering, System (1960s)

In response to the limitations of CB radio, the Federal Communications Commission (FCC) introduced the Class A Citizens Radio Service in the 1960s. This marked a significant step towards a more regulated and organized approach to two-way radio communication for personal use. These early GMRS predecessors utilized tube-type technology, offering wider bandwidth compared to modern systems. However, limitations in power output (around 60 watts) and wider channel spacing (50 kHz) resulted in inefficiencies. While the Class A Citizens Radio Service provided a glimpse into the potential of personal two-way radio communication, its technical shortcomings paved the way for further refinement.

A Turning Point: The Birth of GMRS and the Move Towards Efficiency (1970s-1980s)

The 1970s witnessed a crucial turning point in the evolution of GMRS. Recognizing the growing demand for reliable personal communication and the limitations of the Class A system, the FCC implemented several key changes. Firstly, to address congestion and free up channels for personal users, businesses were gradually transitioned to dedicated

Business Radio Service channels. This move significantly reduced interference and improved overall system efficiency.

Secondly, the FCC standardized power output for GMRS radios, typically around 50 watts. This ensured consistent communication range and helped to minimize signal strength variations. Furthermore, channel spacing was narrowed to 25 kHz, effectively doubling the number of available channels within the allocated spectrum. These changes, coupled with the transition from tube-type to transistorized radios, marked a significant leap forward in terms of portability, ease of use, and overall system effectiveness.

Finally, the late 1980s saw the official christening of the "General Mobile Radio Service," replacing the earlier Class A designation. This shift not only reflected the evolving nature of the technology but also emphasized its focus on mobile communication for personal and recreational purposes.

The Digital Revolution: A New Era for GMRS (1990s-Present)

The turn of the 21st century ushered in a new era for GMRS – the digital revolution. While analog GMRS radios remain the dominant force, digital models are steadily gaining traction. These digital radios offer potential advantages like improved audio quality, stronger signal strength, and the possibility of incorporating more advanced features like encryption or data transmission capabilities.

The transition to digital technology within GMRS, however, is not without its challenges. Digital radios typically come at a higher cost point compared to their analog counterparts. Additionally, compatibility between analog and digital models can be limited, requiring users within a communication group to choose a single standard. Despite these hurdles, the potential benefits of digital technology are undeniable, and its continued development holds promise for the future of GMRS.

The Evolving Landscape: Looking Ahead to the Future of GMRS

The story of GMRS is one of continuous adaptation and improvement. Regulatory bodies like the FCC are constantly evaluating potential advancements, such as the allocation of additional channels or the exploration of advanced data functionalities. Here's a glimpse into some potential areas of future development:

- **Spectrum Allocation and Channel Expansion:** As the demand for GMRS radios grows, the need for more communication channels might become evident. The FCC could potentially explore allocating additional spectrum or implementing techniques like channel sharing to maximize efficiency.
- **Data Capabilities and Interconnectivity:** The future of GMRS might witness a more prominent role for data transmission. Imagine sending basic text messages, sharing GPS coordinates, or even integrating rudimentary internet connectivity for weather updates or map access. These functionalities could enhance the versatility of GMRS, particularly for outdoor enthusiasts or emergency preparedness scenarios.
- **Advanced Features and Interoperability:** Technological advancements could pave the way for the incorporation of more sophisticated features in GMRS radios. Voice encryption for secure communication, improved noise cancellation algorithms, or even integration with other communication platforms are all possibilities on the horizon. Additionally, ensuring seamless interoperability between different GMRS radio brands and models would enhance user experience and flexibility.
- **Regulatory Considerations and User Education:** As GMRS technology evolves, the FCC might need to adapt regulations to ensure responsible use and prevent interference with other radio services. Furthermore, ongoing user education regarding proper licensing procedures, responsible communication practices, and safe operation of GMRS radios remains crucial.

The future of GMRS is brimming with possibilities. By understanding the rich history of innovation and adaptation that brought us to the current state of the technology, we can better appreciate its potential and anticipate the exciting advancements that lie ahead. This journey through time highlights the enduring human desire for reliable communication, a desire that GMRS radios continue to fulfill in a constantly evolving technological landscape.

1.3 Importance and Applications: Unleashing the Potential of GMRS Radios

In today's world, where instant connectivity is often taken for granted, GMRS radios carve out a unique niche. They offer a reliable and independent communication channel, particularly valuable in situations where cell phone service might be patchy or nonexistent. But the importance of GMRS goes beyond mere convenience. Let's peek into the diverse applications that make GMRS radios a valuable tool for a wide range of users.

1.3.1 The Great Outdoors: A Lifeline for Adventure Seekers

For those who crave the thrill of exploration, be it hiking through remote trails, navigating backcountry roads, or embarking on camping adventures, GMRS radios offer an invaluable safety net. Here's how:

- **Enhanced Coordination:** Imagine coordinating a hike with your group or ensuring everyone stays on the same page during a challenging off-road expedition. GMRS radios provide instant communication, allowing you to regroup, share updates, or simply maintain a sense of connection in the wilderness.
- **Emergency Preparedness:** Accidents or unforeseen circumstances can arise in any outdoor setting. GMRS radios equip you with a direct line to call for help in case of emergencies. Whether it's a medical situation, a search and rescue operation, or simply requesting assistance due to a mechanical issue with your vehicle, having a reliable communication tool can make a world of difference.
- **Weather Monitoring:** Many GMRS radios offer access to NOAA weather channels, providing real-time updates on weather conditions in your vicinity. This crucial information allows you to make informed decisions and adjust your plans accordingly, ensuring a safer and more enjoyable outdoor experience.

1.3.2 Business Efficiency: Streamlining On-Site Communication

Beyond the realm of outdoor enthusiasts, GMRS radios play a significant role in streamlining communication for various businesses. Here are some prime examples:

- **Construction Sites:** Imagine a bustling construction site where clear and coordinated communication is paramount. From coordinating between crews to dispatching instructions, GMRS radios provide a dedicated channel for efficient on-site communication, ensuring tasks are completed seamlessly.
- **Security Personnel:** For security guards patrolling a large property or event staff managing crowd control, GMRS radios offer a reliable way to maintain constant communication and share updates. This allows for a more coordinated and efficient response to any situation that might arise.
- **Agriculture and Farming:** Spread across vast landscapes, farms and agricultural operations often require communication between workers in remote locations. GMRS radios bridge the gap, facilitating coordination for tasks, monitoring livestock, or simply ensuring the safety and well-being of personnel on the ground.

1.3.3 Beyond the Expected: Unconventional Uses for GMRS Radios

The versatility of GMRS radios extends beyond these typical applications. Here are some less conventional, yet equally valuable, uses:

- **Neighborhood Watch Groups:** These radios can be a valuable tool for neighborhood watch groups, allowing residents to communicate suspicious activity or coordinate efforts in case of emergencies.
- **Community Events:** Organizers of marathons, cycling races, or other large-scale events can utilize GMRS radios for efficient communication between volunteers, security personnel, and event staff.
- **Disaster Preparedness:** In the unfortunate event of a natural disaster or other large-scale crisis where traditional communication infrastructure might be compromised, GMRS radios can serve as a vital lifeline for community members to share information, coordinate relief efforts, or simply maintain a sense of connection during challenging times.

1.3.4 Unveiling the Potential for Personal Use

Beyond these professional applications, GMRS radios offer valuable benefits for personal use as well:

- **Maintaining Connection on Road Trips:** Long road trips can often suffer from spotty cell service. GMRS radios provide a reliable way to stay connected with your travel companions, ensuring everyone stays on the same page, particularly while navigating unfamiliar territory.
- **Family Communication:** Whether you're enjoying a day at the beach or attending a crowded event, GMRS radios offer peace of mind for families with young children. Having a dedicated communication channel allows you to easily reach your family members in case of separation or simply stay connected throughout the day.
- **Hobbyist Activities:** For enthusiasts of remote control airplanes, drones, or other radio-controlled hobbies, GMRS radios can be used for coordination and communication within designated frequencies, ensuring safe and enjoyable operation of these hobbies.

By understanding these diverse applications, you gain a deeper appreciation for the true potential of GMRS radios. They are more than just walkie-talkies; they are a communication tool that empowers individuals and groups to operate effectively, stay safe, and maintain a sense of connection in various situations. The next chapter will look into the legalities surrounding GMRS radio use, ensuring you operate within the proper regulatory framework.

1.4 Legal Regulations and Licensing Requirements: Navigating the GMRS Landscape

Unleashing the potential of GMRS radios hinges on understanding and adhering to the legal regulations governing their use. Unlike CB radios, GMRS operation necessitates a license issued by the Federal Communications Commission (FCC) in the United States. This licensing framework ensures responsible use of the allocated frequencies and minimizes interference within the GMRS spectrum. Here's a breakdown of the key points to consider:

1.4.1 Eligibility and Application Process:

- **Who Can Apply?** The FCC allows any individual 18 years of age or older who is not a representative of a foreign government to apply for a GMRS license. The application process is relatively straightforward and can be completed online through the FCC's website.
- **Fees and Validity:** The current application fee for a GMRS license sits at $35, and the license itself remains valid for a period of 10 years. Renewal is required before the license expiration date to ensure continued legal operation of your GMRS radio.
- **Sharing the License:** A unique aspect of the GMRS license is that it covers the licensee and all immediate family members residing in the same household, regardless of age. This allows for cost-effective family communication using a single license.

1.4.2 Responsible Use and FCC Guidelines:

- **Understanding the Rules:** The FCC outlines specific regulations governing GMRS radio use within its Part 95 Subpart E of the Code of Federal Regulations. Familiarizing yourself with these regulations is crucial to ensure responsible operation and prevent potential violations.
- **Prohibited Activities:** The FCC strictly prohibits using GMRS radios for any illegal activity, transmitting false or misleading information, music, advertisements, or political campaign messages. The primary purpose of GMRS communication is to facilitate two-way voice communication for personal or business use.
- **Maintaining Proper Etiquette:** While not strictly an FCC regulation, proper communication etiquette goes a long way in ensuring a smooth and enjoyable experience for all GMRS users. This includes being mindful of channel congestion, using clear and concise language, and avoiding excessive chatter that might disrupt others.

1.4.3 Understanding Call Signs and Identification:

- **The Importance of Call Signs:** Each FCC-issued GMRS license comes with a unique call sign. This call sign acts as your identifier on the airwaves, allowing other GMRS users to recognize your transmissions.
- **Transmission Identification:** FCC regulations mandate that GMRS users identify their station with their assigned call sign at the end of each transmission and periodically during extended conversations. This identification helps maintain order on the channels and prevents confusion.

1.4.4 Power Limitations and Channel Sharing:

- **Understanding Power Output:** The FCC establishes specific limitations on the maximum power output permitted for GMRS radios. These limitations help to prevent interference and ensure fair channel access for all users. Mobile and base stations are typically limited to 50 watts, while handheld units have a lower power output to conserve battery life.
- **Sharing the Spectrum:** The GMRS spectrum is allocated to shared use with certain Federal and non-federal entities. While unlikely in most situations, it's important to be aware of the possibility that GMRS channels might be temporarily unavailable due to priority use by these entities.

1.4.5 Consequences of Non-Compliance:

Failing to adhere to FCC regulations can lead to penalties and potential enforcement actions. These can range from warnings and fines to the suspension or revocation of your

GMRS license. Understanding and following the regulations ensures responsible use and protects the integrity of the GMRS system for everyone.

Familiarizing yourself with these legal aspects of GMRS radio operation helps you navigate the communication landscape with confidence. The next section will explore the technical side of the spectrum, explaining how GMRS utilizes specific frequencies and how these allocations work to prevent interference.

1.5 Spectrum Allocation and Frequency Coordination

Imagine a bustling city with designated lanes for different types of vehicles. Cars travel in their designated lanes, while motorcycles might have a separate lane for efficient movement. The GMRS radio world operates on a similar principle, but instead of roads and vehicles, we're dealing with invisible radio waves and communication channels.

1.5.1 Unveiling the Radio Wave Ocean: The Electromagnetic Spectrum

Think of all the wireless communication happening around you – radio broadcasts, Wi-Fi signals, cell phone calls. All these rely on invisible radio waves traveling through the air. The entirety of these radio waves makes up a vast invisible ocean called the electromagnetic spectrum. This spectrum is like a giant highway with different sections allocated for various purposes, just like the lanes in our city example.

1.5.2 Finding Your Lane: The GMRS Frequency Band

Regulatory bodies like the FCC act as the traffic controllers of this radio wave highway. They designate specific portions of the spectrum for different uses. For GMRS radios, a slice of the Ultra High Frequency (UHF) band is allocated, specifically between 462 MHz and 467 MHz. Think of this band as your dedicated lane on the radio wave highway, allowing GMRS users to communicate without interfering with other types of radio traffic.

1.5.3 Channel Surfing: Keeping Conversations Separate

Now, within this GMRS lane, the FCC further divides things up into individual channels. Imagine these channels as designated conversation zones within your lane. Just like lanes on a highway are spaced apart to prevent collisions, GMRS channels are spaced out (typically by 12.5 kHz or 25 kHz) to ensure your conversation doesn't bleed into someone else's. Most GMRS radios come equipped with a selection of these channels, allowing you to choose a clear one for your communication needs. It's like selecting a radio station – you tune in to the specific channel where your conversation is happening.

1.5.4 Boosting Your Signal: The Role of Repeaters

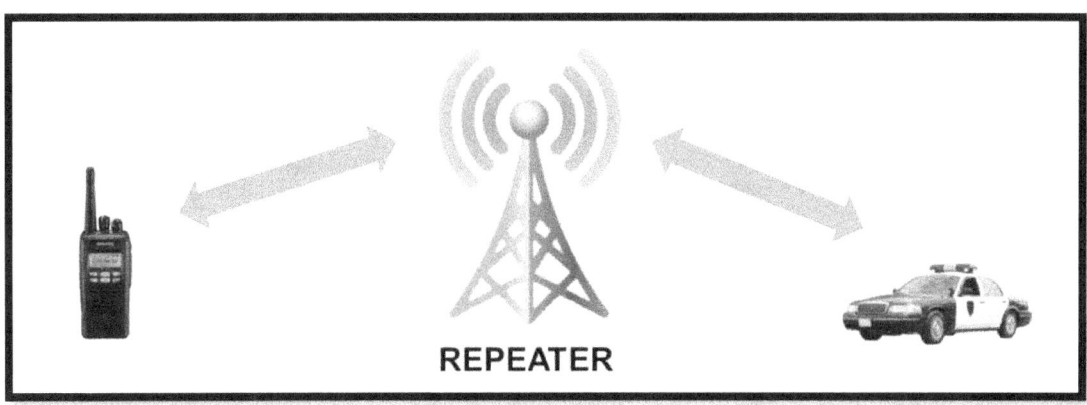

REPEATER

While GMRS radios offer decent range in ideal conditions (think open fields), limitations arise due to obstacles like mountains or distance. Here's where repeaters come in – they act like signal boosters strategically placed on high points. Imagine a friend shouting your message from a hilltop to someone further away – that's essentially what a repeater does. It receives your weak GMRS signal, amplifies it, and then retransmits it on the same or a different frequency, effectively extending the range of your communication. However, using repeaters often requires additional licensing and following specific regulations, so it's not always the simplest option for beginners.

1.5.5 The Future of the GMRS Lane: Keeping Up with Demand

Just like a busy city might need to expand its road network to accommodate more traffic, the FCC might need to adjust spectrum allocation for GMRS in the future. With the growing popularity of GMRS radios, the demand for clear communication channels might increase. The FCC could explore options like allocating additional channels within the existing GMRS band or even implementing techniques for sharing channels more efficiently.

1.5.6 Choosing Your Ride: Selecting the Right GMRS Radio

Understanding these technical concepts empowers you to make informed decisions when purchasing a GMRS radio. Here are some factors to consider:

- **Number of Channels:** Think of this as the number of radio stations your radio can tune into. More channels offer more flexibility in finding a clear one for your communication needs.
- **Repeater Compatibility:** If you anticipate using repeaters for extended range, ensure your chosen radio has this functionality.
- **Radio Power Output:** Higher power output radios can potentially reach longer distances, but regulations limit this power, so focus on finding a radio that meets your typical communication needs.

Understanding the basics of spectrum allocation and frequency coordination puts you well on your way to navigating the world of GMRS radios with confidence.

Summary

This chapter provided a thorough introduction to GMRS radios, arming you with essential knowledge to navigate this thrilling domain of two-way communication. We examined the fundamental operations of GMRS radios, comprehending their application for brief voice exchanges on specified UHF frequencies. You acquired knowledge about the requirement for FCC licensing in the United States and deepened your understanding of the technical facets of radio waves, spectrum allocation, and frequency coordination.

We traced the evolution of GMRS from its origins as early Citizen Band Radio Service systems to its current standardized and efficient form. The potential advantages of digital GMRS technology were discussed, emphasizing its potential for enhanced audio quality and future developments.

A substantial part of the chapter was dedicated to the various applications of GMRS radios. From enhancing safety for outdoor enthusiasts to facilitating communication for businesses such as construction crews or security personnel, GMRS radios offer a dependable and independent communication pathway in a range of scenarios. We also explored unconventional uses for GMRS radios, including neighborhood watch groups or event management.

Understanding the legal regulations governing GMRS radio use is essential. The chapter detailed the FCC licensing process, eligibility criteria, and responsible communication practices. You gained insights into call signs, proper identification procedures, and the importance of complying with FCC regulations to ensure a seamless and enjoyable experience for all users.

Furthermore, we explored the technical aspects of spectrum allocation and frequency coordination. By understanding how GMRS employs specific UHF frequencies and how channels are arranged to avoid interference, you developed a deeper appreciation for the complex mechanisms of the GMRS system.

This chapter equipped you with a solid foundation for exploring the practicalities of GMRS radios in the subsequent chapters. We'll explore selecting the appropriate equipment, proper communication etiquette, and all the exciting opportunities that GMRS radios offer for reliable and effective communication in today's world.

Review Questions

1. This chapter explored the concept of spectrum allocation and frequency coordination for GMRS radios. Imagine a busy highway with multiple lanes for different types of vehicles. How does this analogy relate to the concept of spectrum allocation for GMRS radios within the electromagnetic spectrum?

2. The chapter emphasized the importance of adhering to FCC regulations for GMRS radio use. Briefly explain the two key points covered regarding:

   - A) Who is eligible to apply for a GMRS license?
   - B) What are some of the prohibited activities when using a GMRS radio?

3. Imagine explaining GMRS radio channels to a friend who has no prior experience with radios. How would you explain, in simple terms, the concept

of channels and their importance in GMRS communication, using an analogy if possible?

CHAPTER 2

UNDERSTANDING GMRS RADIO SYSTEMS

Having laid a solid foundation in the functionalities, applications, and legal aspects of GMRS radios in Chapter 1, we're now prepared to explore the technical aspects. This chapter acts as your guide to understanding GMRS radio systems, exploring the essential components that enable these radios to function and the various features that augment their capabilities. Through this exploration, you'll acquire valuable knowledge to select the appropriate GMRS radio for your specific needs and maximize its effectiveness in various communication scenarios.

We'll start by examining the core components of a GMRS radio, understanding the role each element plays in transmitting and receiving signals. Next, we'll differentiate between handheld and mobile GMRS radios, highlighting the advantages and considerations for each type. The chapter will explore the world of antennas and accessories, detailing how these external elements can optimize your communication range and functionality.

As technology progresses, so do GMRS radios. We'll explore the exciting world of advanced signal processing technologies, specifically focusing on Digital Signal Processing (DSP) and its impact on audio quality and performance. Furthermore, we'll examine noise reduction and voice enhancement features, understanding how these functionalities can significantly improve clarity and intelligibility, particularly in challenging communication environments.

Finally, the chapter will conclude by exploring the concept of RF (Radio Frequency) propagation and the impact of terrain on signal transmission. By understanding how radio waves travel and the influence of obstacles like mountains or buildings, you'll be well-equipped to choose the right locations and strategies for optimal communication with your GMRS radio.

Throughout this chapter, we'll strive to present the technical concepts in a clear and concise manner, utilizing analogies and practical examples to ensure understanding for beginners and enthusiasts alike. So, buckle up and prepare to unravel the fascinating world of GMRS radio systems!

2.1 GMRS Radio Components and Features: Unveiling the Building Blocks

Imagine a walkie-talkie – a simple two-way communication device used for short-range conversations. Now, picture that walkie-talkie on steroids, packed with additional features and functionalities – that's essentially what a GMRS radio is. But just like any sophisticated tool, GMRS radios are comprised of various internal components working together seamlessly to transmit and receive signals. Let's break down these essential building blocks and understand their roles in ensuring effective GMRS communication:

- **Transceiver:** Consider the transceiver the heart of your GMRS radio. This crucial component acts as a two-way translator, converting electrical signals representing your voice into radio waves for transmission and vice versa. Imagine speaking into a microphone – the transceiver takes those sound waves and transforms them into electrical signals. During receive mode, the process flips, with the transceiver converting incoming radio waves back into electrical signals that your speaker can translate into audible sound.

- **Transmit and Receive Antennas:** Think of antennas as the ears and mouth of your GMRS radio. The transmit antenna captures the radio waves generated by the transceiver and radiates them outwards, essentially carrying your voice as electrical signals converted into radio waves. Conversely, the receive antenna acts as the ear, picking up incoming radio waves from other GMRS users and feeding them back to the transceiver for processing and conversion into audible sound. The size and design of the antenna significantly impact your communication range, so we'll go deeper into this aspect later in the chapter.

- **Battery:** Powering the entire operation is the battery. Without a reliable power source, your GMRS radio becomes nothing more than a paperweight. When selecting a GMRS radio, consider factors like battery life and type (rechargeable versus disposable) to ensure it aligns with your intended use.

- **Speaker and Microphone:** These components are fairly self-explanatory. The speaker allows you to hear incoming voice transmissions from other GMRS users, while the microphone picks up your voice and converts it into electrical signals for transmission. Some GMRS radios offer features like adjustable speaker volume or noise-canceling microphones for enhanced communication clarity.

- **Channel Selector and Display:** Imagine a television with multiple channels – a GMRS radio functions similarly. The channel selector allows you to choose the specific frequency (channel) on which you want to transmit and receive communication. The display unit typically shows the selected channel, battery level, signal strength, and sometimes even additional information depending on the radio model.
- **Push-to-Talk (PTT) Button:** This button acts as the trigger for your communication. When you press and hold the PTT button, the radio switches from receive mode to transmit mode, allowing your voice to be broadcast through the antenna. Releasing the PTT button switches the radio back to receive mode, enabling you to hear incoming transmissions from others.
- **Optional Features:** Beyond these core components, some GMRS radios offer additional features that enhance their functionality. These might include features like scanning capabilities to search for active channels, roger beep tones to indicate the end of your transmission, or even integrated VOX (Voice Operated Transmit) functionality, allowing hands-free communication by transmitting your voice when picked up by the microphone.

This knowledge empowers you to make informed decisions when selecting a radio that meets your specific needs and effectively utilize its features for clear and reliable communication. The next section will explore the two main types of GMRS radios available – handheld and mobile units – highlighting their advantages and considerations for different use cases.

2.2 Types of GMRS Radios: Handheld vs. Mobile Units – Tailoring Communication to Your Needs

The world of GMRS radios offers two main categories to choose from – handheld and mobile units. Understanding the distinct advantages and considerations of each type is crucial for selecting the right radio that aligns with your intended use. Here's a detailed breakdown to guide your decision-making process:

2.2.1 Handheld GMRS Radios: The Grab-and-Go Champions of Communication

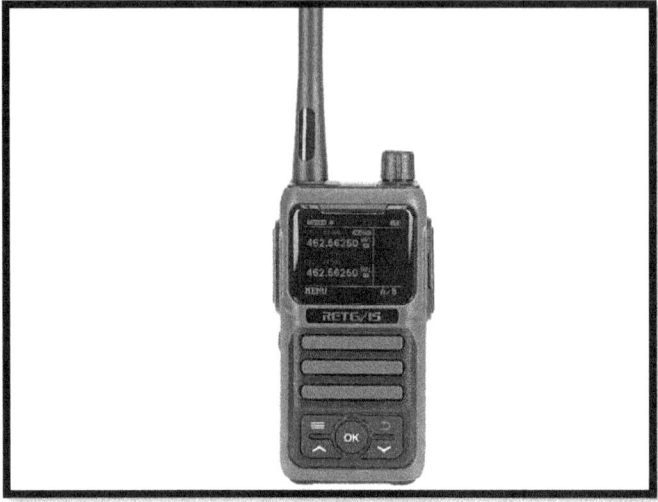

Imagine a compact, lightweight walkie-talkie you can easily carry around. That's the essence of a handheld GMRS radio. These radios are ideal for situations where portability and ease of use are paramount. Let's explore the key strengths that make handheld GMRS radios so popular:

- **Unmatched Portability:** Their compact size and lightweight design allow you to effortlessly carry them in your hand, clip them onto a belt, or store them in a backpack. This makes them perfect for outdoor activities like hiking, camping, or exploring remote areas where on-the-go communication is essential.
- **Convenience and Ease of Use:** Handheld GMRS radios are generally user-friendly, with simple controls and interfaces. Most models require minimal setup, allowing you to power them on, select a channel, and start communicating quickly. This ease of use makes them ideal for beginners or anyone who needs a straightforward communication solution.
- **Versatility in Various Applications:** Beyond outdoor adventures, handheld GMRS radios find applications in various scenarios. Construction crews can utilize them for on-site coordination, event organizers can leverage them for streamlined communication between staff, or even neighborhood watch groups can benefit from their portability for maintaining a sense of security.

- **Power Efficiency:** Handheld GMRS radios are designed to be power-efficient, often utilizing features like automatic power-saving modes to extend battery life. This is particularly advantageous in situations where access to power sources might be limited, ensuring your radio remains operational throughout your adventure.

However, alongside these strengths, it's important to consider the limitations of handheld GMRS radios:

- **Limited Communication Range:** Due to their compact size and lower antenna placement, handheld radios typically offer a shorter communication range compared to mobile units. The range can be further impacted by terrain and obstacles like hills or buildings.
- **Lower Power Output:** Regulations limit the maximum power output of handheld GMRS radios compared to mobile units. This translates to a shorter overall communication range, especially in challenging conditions.
- **Audio Quality:** While adequate for most basic communication needs, handheld radios might have limitations in terms of audio fidelity compared to mobile units with larger speakers and microphones. Background noise or less-than-ideal environmental conditions can affect audio clarity on handheld radios.

2.2.2 Mobile GMRS Radios: Powerhouses of Communication for Vehicles and Fixed Locations

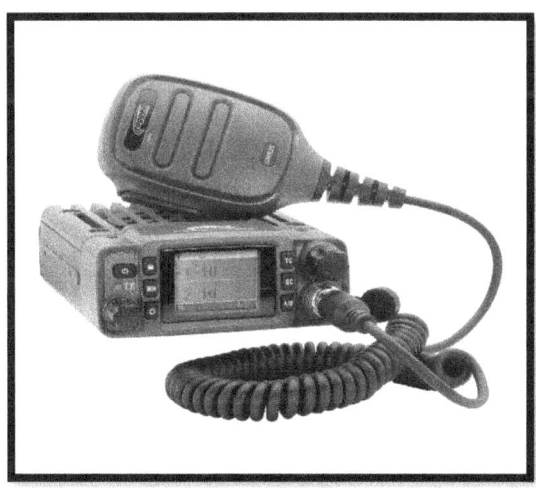

Mobile GMRS radios are designed for installation in vehicles or fixed locations like base stations. They offer a different set of advantages compared to their handheld counterparts:

- **Extended Communication Range:** Mobile GMRS radios typically boast higher power output capabilities and often utilize larger, more powerful antennas strategically positioned on vehicles or mounted at higher locations. This translates to a significantly extended communication range, making them ideal for covering wider areas or overcoming challenging terrain.
- **Superior Audio Quality:** Mobile GMRS radios often come equipped with larger speakers and microphones compared to handheld units. This translates to clearer, more intelligible audio, especially beneficial in noisy environments or for longer communication sessions.
- **Versatility with External Accessories:** Mobile GMRS radios offer greater flexibility for customization. External antennas with higher gain can be installed to further extend range, and external speakers or microphones can be connected for enhanced audio quality in specific situations.
- **Integration with Other Communication Systems:** Some mobile GMRS radios offer features like Bluetooth connectivity, allowing them to connect to headsets or integrate with other communication systems in vehicles, such as CB radios or ham radio setups.

However, mobile GMRS radios also come with some limitations to consider:

- **Reduced Portability:** Mobile GMRS radios are designed for installation in vehicles or fixed locations, limiting their portability compared to handheld units. They are not ideal for situations where you need to move around freely while communicating.
- **Complexity of Installation:** While typically straightforward, installing a mobile GMRS radio might require some technical knowledge or professional assistance, especially when connecting external antennas or integrating them with other communication systems in a vehicle.
- **Power Consumption:** Mobile GMRS radios typically draw more power compared to handheld units. This might be a consideration if you plan to operate the radio

for extended periods using a vehicle's battery, especially if the alternator capacity is limited.

Choosing the Right Radio: Matching Your Needs to the Perfect Fit

The ideal GMRS radio for you depends on your specific needs. Here are some key factors to consider when making your decision:

- **Primary Use Case:** Identify the primary purpose for your GMRS radio. If portability and ease of use are paramount for outdoor adventures, a handheld unit might be the perfect choice. For extended range communication in a vehicle or fixed location, a mobile GMRS radio offers significant advantages.
- **Communication Range Requirements:** Evaluate the typical distances you expect to cover during communication. If you primarily operate in close proximity to others using GMRS radios, a handheld unit might suffice. However, for covering larger distances or overcoming challenging terrain, a mobile GMRS radio with its extended range capabilities becomes more suitable.
- **Audio Quality Priorities:** Consider the importance of clear audio in your communication scenarios. If background noise is a concern, or you anticipate longer communication sessions, the superior audio quality of mobile GMRS radios might be a deciding factor.
- **Budget Considerations:** GMRS radios come in a range of prices depending on features and functionalities. Handheld units are generally more affordable than mobile units. Set a realistic budget and prioritize the features most important to your needs.
- **Future Needs and Upgradability:** Think about your potential future needs. If you foresee expanding your use of GMRS radios, consider a mobile unit with the potential for customization through external antennas or accessories.

By carefully considering these factors and the strengths and limitations of handheld and mobile GMRS radios, you'll be well-equipped to select the perfect fit for your communication needs. In the next section, we look into the world of antennas and accessories, exploring how these external elements can further enhance the capabilities of your chosen GMRS radio.

2.3 Antennas and Accessories: Optimizing Your GMRS Radio's Performance

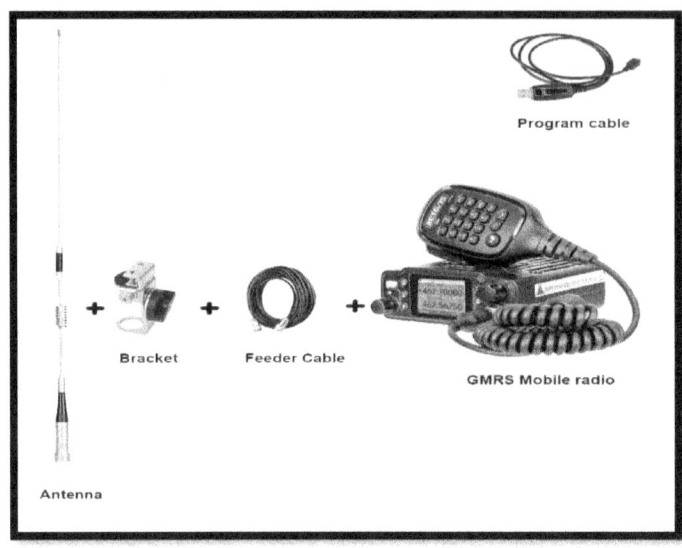

Imagine a car with high-performance tires – they can significantly enhance the vehicle's handling and grip. Similarly, antennas and accessories can play a crucial role in optimizing the performance of your GMRS radio. Let's explore this exciting world of external elements and how they can elevate your communication experience:

2.3.1 Antennas: The Invisible Reach of Your GMRS Radio

The antenna on your GMRS radio acts as the bridge between the electrical signals carrying your voice and the radio waves that transmit them over distances. Just like a flag waving in the wind, the antenna captures and transmits these radio waves. Understanding different antenna types and their functionalities empowers you to choose the right one for maximizing your communication range and effectiveness.

- **Stock Antennas:** Most GMRS radios come equipped with a basic "rubber duck" antenna – a short, flexible antenna typically integrated into the radio itself. While convenient for portability, these antennas offer a limited communication range due to their size and design.
- **Upgraded Handheld Antennas:** For handheld GMRS radios, a wide variety of aftermarket antennas are available. These antennas can be detachable and offer

various lengths and styles. Generally, a longer antenna translates to a potentially increased communication range. However, factors like antenna gain (the ability to focus the radio waves in a specific direction) and local regulations regarding antenna height also play a role.

- **Mobile Antennas:** Mobile GMRS radios typically utilize external antennas mounted on the vehicle. These antennas come in various styles, including:
    - o **J-Pole Antennas:** A popular choice due to their compact size and omnidirectional radiation pattern (meaning they transmit and receive signals equally in all directions).
    - o **Base Station Antennas:** Designed for fixed locations like homes or businesses, these antennas offer superior range due to their height and directional capabilities. They can be mounted on rooftops or other high points for optimal performance.
- **Factors to Consider When Choosing an Antenna:**
    - o **Frequency:** Ensure the antenna is designed for the specific GMRS frequency range (typically 462 MHz to 467 MHz).
    - o **Gain:** Higher gain antennas offer a potentially increased range but might come with a trade-off in terms of omnidirectional coverage.
    - o **Length:** While generally, longer antennas offer better range, consider the practical limitations of size and portability, especially for handheld units.
    - o **Local Regulations:** Some areas might have restrictions on antenna height or type. Always check local regulations before installing an external antenna.

2.3.2 Essential Accessories for Enhanced Functionality

Beyond antennas, a variety of accessories can further enhance the performance and usability of your GMRS radio:

- **Headsets and Earpieces:** These offer hands-free communication, particularly beneficial for situations where you need to keep your hands free, such as driving or performing tasks while communicating. They also improve audio clarity, especially in noisy environments.

- **Batteries and Chargers:** Investing in spare batteries or a reliable external charger ensures you don't run out of power during crucial communication moments. Consider the battery type (rechargeable versus disposable) and choose based on your usage patterns and access to charging options.
- **Belt Clips and Holsters:** These accessories provide convenient carrying options for handheld GMRS radios, keeping them readily accessible during outdoor activities or on-the-go communication.
- **Speaker Mics:** These combine a speaker and microphone into a single unit, offering improved audio quality and hands-free communication capabilities, particularly useful for mobile GMRS radio setups.
- **Scanning Antennas:** These specialized antennas can be used for handheld radios to scan for active channels in your area, helping you identify channels with ongoing communication.

2.3 Antennas and Accessories: Unveiling the Secret Weapons of Your GMRS Radio

Imagine a seasoned archer pulling back the string of a bow. The effectiveness of their shot relies not just on their skill, but also on the quality of the arrow. Similarly, the performance of your GMRS radio hinges not only on its internal components but also on the external elements that extend its reach and functionality. Antennas and accessories act as the secret weapons of your GMRS radio, fine-tuning its communication capabilities and optimizing your overall experience. Let's embark on a detailed exploration of these crucial elements, empowering you to transform your GMRS radio into a communication powerhouse.

2.3.1 Antennas: The Invisible Lifeline of Your GMRS Radio

Think of an antenna as a bridge connecting the electrical signals carrying your voice within the radio to the invisible world of radio waves that transmit them over distances. Just like a conductor's baton translating musical notes into sweeping gestures for the orchestra, the antenna transforms the electrical signals into radio waves and vice versa. Understanding the different types of antennas and their functionalities equips you to select the perfect fit for maximizing your communication range and effectiveness.

- **Stock Antennas: The Built-in Convenience with Limitations**

Most GMRS radios come equipped with a basic "rubber duck" antenna. Imagine a short, flexible antenna typically integrated into the body of the radio itself. While these antennas offer the undeniable advantage of convenience, their compact size translates to a limited communication range. The physics of radio wave propagation dictates that longer antennas generally offer better performance. The shorter length of the stock antenna restricts its ability to efficiently capture and transmit radio waves, limiting the distance your voice can travel.

- **Upgraded Handheld Antennas: Expanding Your Reach on the Go**

For handheld GMRS radios, a vast world of aftermarket antennas awaits. These antennas are typically detachable, allowing you to customize your communication setup based on your needs. Imagine replacing the stock antenna with a longer, more robust version. This seemingly simple change can significantly enhance your communication range. Aftermarket antennas come in various lengths and styles, each with its own advantages:

* **Extended Length Antennas:** As mentioned earlier, the length of the antenna directly impacts its ability to capture and transmit radio waves. A longer antenna, like a metaphorical megaphone for radio waves, allows for more efficient transmission and reception, potentially extending your communication range. However, there's a trade-off to consider. While a meter-long antenna might offer impressive range, its practicality for on-the-go use becomes limited. Finding the right balance between increased range and portability is crucial for handheld antennas.

* **Gain Antennas:** Beyond length, another factor influencing antenna performance is gain. Imagine focusing a beam of light with a magnifying glass – that's essentially what a gain antenna does for radio waves. By concentrating the radio waves in a specific direction, gain antennas offer potentially increased range, particularly for long-distance communication. However, there's a catch. Gain antennas often come with a trade-off in terms of omnidirectional coverage. A high-gain antenna might excel at transmitting in a specific direction, but it might compromise its ability to receive signals equally from all directions. Understanding your communication needs is key – if you primarily

communicate with someone in a known location, a high-gain antenna could be ideal. But if you need to maintain awareness of communication happening around you on different channels, an omnidirectional antenna might be a better choice.

- **Mobile Antennas: Reaching Out from Your Vehicle**

Mobile GMRS radios typically utilize external antennas mounted on the vehicle itself. These antennas come in various styles, each catering to specific needs:

* **J-Pole Antennas:** A popular choice for mobile GMRS setups due to their compact size and versatility. Imagine a J-shaped metal rod with the antenna element at the top. J-Pole antennas offer an omnidirectional radiation pattern, meaning they transmit and receive signals equally in all directions. This makes them ideal for situations where you need to maintain 360-degree communication awareness. While not the most powerful option in terms of range, J-Pole antennas provide a good balance between size, performance, and ease of use for mobile applications.

* **Base Station Antennas:** Designed for fixed locations like homes or businesses, base station antennas offer superior range due to their height and directional capabilities. Imagine a tall vertical mast with an antenna element at the top. These antennas can be mounted on rooftops or other high points, leveraging their height to improve communication range significantly. Base station antennas often come with directional capabilities, allowing you to focus your transmission and reception in specific directions. This can be particularly beneficial for long-distance communication or overcoming challenging terrain that might obstruct radio waves in certain directions.

Choosing the Right Antenna

Selecting the perfect antenna for your GMRS radio requires careful consideration of several factors:

- **Frequency:** Ensure the antenna is designed for the specific GMRS frequency range, typically 462 MHz to 467 MHz. Using an antenna outside this range will significantly compromise its effectiveness. Imagine trying to use a screwdriver as a hammer – it might work in a pinch, but it's not the optimal tool for the job. Similarly, an antenna designed for a different frequency range won't efficiently capture and transmit GMRS radio waves.

- **Gain:** As discussed earlier, gain antennas offer potentially increased range but sacrifice some omnidirectional coverage. Consider your communication needs. If you primarily communicate with someone in a known location, a high-gain antenna could be ideal. But if you need to maintain awareness of communication happening around you on different channels, an omnidirectional antenna might be a better choice. Finding the right balance between range and coverage is crucial.
- **Length:** While generally, longer antennas offer better range, consider the practical limitations of size and portability, especially for handheld units. A two-meter antenna on a handheld radio might offer impressive range, but it wouldn't be very practical for carrying around. For handheld antennas, striking a balance between increased range and portability is essential.
- **Local Regulations:** Some areas might have restrictions on antenna height or type. Always check local regulations before installing an external antenna on your vehicle or at a fixed location. Violating these regulations could result in fines or even confiscation of your antenna.

Beyond the Antenna: Essential Accessories for Enhanced Functionality

While antennas play a crucial role in extending your communication reach, a variety of accessories can further enhance the usability and performance of your GMRS radio:

- **Headsets and Earpieces:** These accessories are a game-changer for situations where you need to keep your hands free. Imagine driving down a bumpy road while trying to hold the radio and communicate – not an ideal scenario. Headsets and earpieces allow for hands-free communication, improving safety and convenience. They also enhance audio clarity, especially in noisy environments like a moving vehicle or a crowded campsite.
- **Batteries and Chargers:** Running out of power during a critical communication moment can be frustrating. Investing in spare batteries or a reliable external charger ensures your GMRS radio remains operational when you need it most. Consider the battery type (rechargeable versus disposable) and choose based on

your usage patterns and access to charging options. Imagine being on a remote camping trip and relying solely on disposable batteries – spare batteries or a solar charger become essential companions.

- **Belt Clips and Holsters:** These accessories provide convenient carrying options for handheld GMRS radios. Imagine having your radio readily accessible on your belt while exploring a new trail or coordinating tasks during a busy event. Belt clips and holsters keep your radio within easy reach, ensuring you don't miss important communication.
- **Speaker Mics:** These combine a speaker and microphone into a single unit, offering improved audio quality and hands-free communication capabilities, particularly useful for mobile GMRS radio setups. Imagine having a dedicated speaker and microphone mounted in your vehicle, transforming your GMRS radio into a robust communication system for on-the-road coordination.
- **Scanning Antennas:** These specialized antennas can be used for handheld radios to scan for active channels in your area. Imagine attending a large event where multiple groups might be using GMRS radios. A scanning antenna helps you identify channels with ongoing communication, allowing you to connect with the relevant group.

The power to customize your GMRS radio setup to perfectly match your communication needs lies in your understanding of the different types of antennas, their functionalities, and the various accessories available as discussed in this section. In the next section, we look into the fascinating world of advanced signal processing technologies, exploring how they further elevate the communication experience with GMRS radios.

2.4 Advanced Signal Processing Technologies: Unveiling the Hidden Potential of Your GMRS Radio

Imagine two walkie-talkies connected over a long distance, but the crackling static and background noise make communication nearly impossible. Now, picture those same walkie-talkies equipped with advanced technology that filters out the noise and delivers crystal-clear audio. That's the magic of advanced signal processing technologies in GMRS radios. This section looks into these fascinating functionalities, exploring how they refine and enhance your communication experience.

2.4.1 Digital Signal Processing (DSP): The Brains Behind the Enhancement

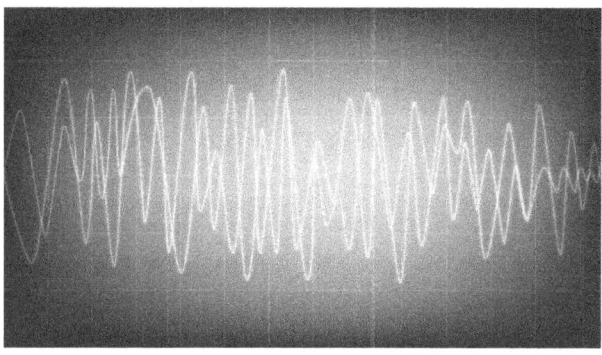

At the heart of advanced signal processing lies a powerful technology called Digital Signal Processing (DSP). Think of DSP as a sophisticated digital toolbox with various functionalities that manipulate the electrical signals carrying your voice before transmission and after reception. By applying these functionalities, DSP significantly improves the overall quality, clarity, and intelligibility of your communication on a GMRS radio.

- **Noise Reduction:** One of the most noticeable benefits of DSP is its ability to reduce unwanted noise. Imagine a roaring campfire crackling in the background while you're trying to communicate with your friend on a GMRS radio. DSP algorithms can identify and filter out this background noise, significantly improving the clarity of your voice transmission.
- **Speech Compression:** DSP can also employ speech compression techniques. Imagine a data highway with limited lanes – compressing speech data allows for more efficient transmission within the available bandwidth of a GMRS radio channel. This translates to potentially clearer audio and less signal distortion, especially in challenging communication environments.
- **Error Correction:** Radio waves can sometimes be affected by interference or atmospheric conditions, leading to errors in the transmitted signal. DSP can incorporate error correction functionalities. Imagine having a scrambled message with missing pieces – error correction algorithms work to reconstruct the original signal, ensuring your voice is delivered clearly and accurately to the receiver.

- **Automatic Gain Control (AGC):** Fluctuations in signal strength can occur during communication, especially with handheld radios or in varying terrain. Imagine someone whispering followed by someone shouting on the radio – dramatic changes in volume can be disruptive. DSP with AGC features can automatically adjust the gain (signal strength) to maintain consistent audio levels, ensuring comfortable listening for the receiver.

2.4.2 Unveiling Additional Enhancements: A Range of Technologies

Beyond the core functionalities of DSP, various other advanced signal processing technologies can further elevate your GMRS radio communication:

- **Voice Activation (VOX):** Imagine being hands-free while using your GMRS radio. VOX technology detects your voice and automatically transmits when you speak, eliminating the need to press a push-to-talk (PTT) button every time. This can be particularly beneficial for situations where you need to keep your hands free, such as during outdoor activities or operating a vehicle.
- **Digital Squelch:** Imagine trying to listen to a faint radio signal amidst a constant hiss or static noise. Digital squelch acts as a sophisticated noise gate. By setting a specific noise threshold, it essentially mutes the speaker when the signal strength falls below that level. This eliminates the constant background noise and allows you to hear only clear transmissions exceeding the squelch threshold.
- **Pre-Emphasis and De-Emphasis:** These functionalities work together to improve audio quality at higher frequencies, which are more susceptible to attenuation (weakening) over longer distances. Imagine high-pitched sounds becoming muffled during communication – pre-emphasis boosts these frequencies slightly before transmission, and the receiving radio applies de-emphasis to restore the original balance, resulting in more natural-sounding audio.

You gain a deeper appreciation for the capabilities of your GMRS radio by understanding these advanced signal processing technologies and their functionalities. These advancements are not just fancy features; they translate into real-world benefits, ensuring clear, reliable, and enjoyable communication experiences, regardless of the environment. The next section will explore another crucial aspect influencing communication effectiveness – RF (Radio Frequency) propagation and the impact of terrain on signal transmission.

2.5 Understanding RF Propagation: The Invisible Landscape Shaping Your GMRS Communication

Imagine shouting across a vast canyon. The sound waves carrying your voice travel outwards, but the canyon walls act as barriers, potentially weakening or even blocking the sound from reaching the other side. Similarly, radio waves used in GMRS communication are susceptible to the influence of the environment they travel through. This section goes into the fascinating world of RF (Radio Frequency) propagation, exploring how terrain and other factors impact the reach and effectiveness of your GMRS radio.

2.5.1 Beyond Line-of-Sight Communication: Understanding Propagation Mechanisms

Unlike a laser beam traveling in a straight line, radio waves can travel over long distances by bending and reflecting off objects in their path. Imagine throwing a pebble into a still pond – ripples emanate outwards in all directions. Radio waves behave similarly, propagating through various mechanisms:

- **Line-of-Sight Propagation:** In ideal scenarios with a clear line of sight between transmitter and receiver, radio waves travel directly from the antenna of one radio to the antenna of the other. This is the most efficient mode of propagation, offering the maximum communication range. However, achieving a true line of sight is not always possible, especially over long distances or in built-up environments.

- **Ground Wave Propagation:** Radio waves can also travel along the surface of the earth, hugging the ground like a wave following the coastline. This propagation mode is particularly effective for lower frequencies, including those used in GMRS radio. However, the ground acts as a conductor, absorbing some of the signal energy, leading to a gradual weakening of the signal strength with increasing distance.

- **Skywave Propagation:** At higher frequencies, radio waves can reflect off layers of the ionosphere, a region of the upper atmosphere charged with electrically charged particles. Imagine a radio wave bouncing off a giant mirror in the sky. This reflection allows for communication over vast distances, but it's a less reliable mode of propagation for GMRS radios due to the unpredictable nature of the ionosphere.

2.5.2 Terrain and the Impact on Signal Strength

The type of terrain you're communicating through significantly impacts the strength and reach of your GMRS radio signal. Imagine radio waves trying to navigate a dense forest compared to a flat, open field. Here's how different terrains can affect communication:

- **Flat, Open Terrain:** This is the ideal scenario for GMRS radio communication. With minimal obstacles in the path of the radio waves, they can travel farther with minimal signal degradation. Imagine a clear line of sight between two points – communication range is maximized in such conditions.
- **Hilly or Mountainous Terrain:** Hills and mountains act as barriers, potentially weakening or even blocking the signal. Imagine radio waves struggling to climb over a mountain – communication range can be significantly reduced in such scenarios. Depending on the severity of the terrain and the positioning of radios, alternative communication strategies might be necessary.
- **Urban Environments:** Buildings, skyscrapers, and other structures in urban areas can cause signal scattering and absorption. Imagine radio waves bouncing off multiple surfaces in a city – this can lead to a weakened and potentially distorted signal. The range of GMRS radios might be reduced in urban environments compared to open areas.

2.5.3 Overcoming Challenges: Strategies for Effective Communication

Understanding how terrain and other factors influence RF propagation empowers you to develop strategies for effective communication with your GMRS radio:

- **Utilize Higher Ground:** If possible, try to communicate from higher points like hills or rooftops. This provides a clearer line of sight for the radio waves and can significantly improve communication range, especially in areas with challenging terrain.

- **Plan Your Communication Routes:** When venturing into areas with potentially difficult terrain, plan your communication routes beforehand. Identify high points or open areas that might offer better signal transmission. Having a backup plan or alternative communication methods can be beneficial in case of unforeseen signal limitations.
- **Invest in Higher Gain Antennas:** As discussed earlier, antennas with higher gain can focus the radio waves in a specific direction, potentially overcoming signal-weakening obstacles like hills or buildings. Consider using a higher gain antenna if you anticipate communicating in areas with challenging terrain.
- **Respect the Limitations:** While GMRS radios offer a decent communication range, it's important to respect their limitations. Don't rely solely on GMRS radio communication for critical situations where long-distance or absolutely reliable communication is paramount. Consider alternative communication methods as backups in case GMRS radio signals become weak or unavailable.

With the understanding the concepts of RF propagation and the impact of terrain, you gain the knowledge to navigate the invisible landscape that shapes your GMRS radio communication.

Summary

This chapter transformed you from a beginner to a confident GMRS radio user. We started by exploring the radio's interior – the transceiver, antennas, battery, and the tools for speaking and listening (speaker and microphone). We learned how channels and the push-to-talk button keep communication organized.

Next, we compared handheld and mobile GMRS radios, highlighting the strengths and weaknesses of each to help you choose the perfect fit for your adventures. We then ventured into the world of antennas and accessories, explaining how different types of antennas can extend your communication range. Headsets, batteries, and speaker mics were introduced as tools to enhance your radio's usability.

But a powerful GMRS radio is more than just hardware. We explored the magic of signal processing technology, revealing how it reduces noise, clarifies voices, and ensures messages arrive clearly. We even touched on features like voice activation and automatic volume control for a smoother communication experience.

The chapter didn't stop there. We journeyed into the fascinating world of radio waves, unveiling how mountains, buildings, and even the Earth itself can affect your signal. Learning these invisible forces allows you to develop strategies for better communication, like using higher ground or alternative communication methods when needed.

By equipping you with this knowledge, this chapter empowers you to confidently operate your GMRS radio and enjoy clear, reliable communication, wherever your adventures take you. The next chapter will focus on the legal aspects of using GMRS radios, ensuring you operate responsibly within the set guidelines.

Review Questions

1. The chapter explored two main types of GMRS radios. Can you identify the key factors to consider when choosing between a handheld and a mobile GMRS radio?
2. GMRS radios boast more than just channels and a PTT button. Describe two ways advanced signal processing technology enhances your communication experience with a GMRS radio.
3. Radio waves travel differently than sound waves. Explain how understanding RF propagation, like the impact of terrain, can help you improve your GMRS radio communication strategies.

CHAPTER 3

MASTERING THE LANGUAGE OF GMRS RADIOS – FREQUENCIES, SIGNALS, AND PROTOCOLS

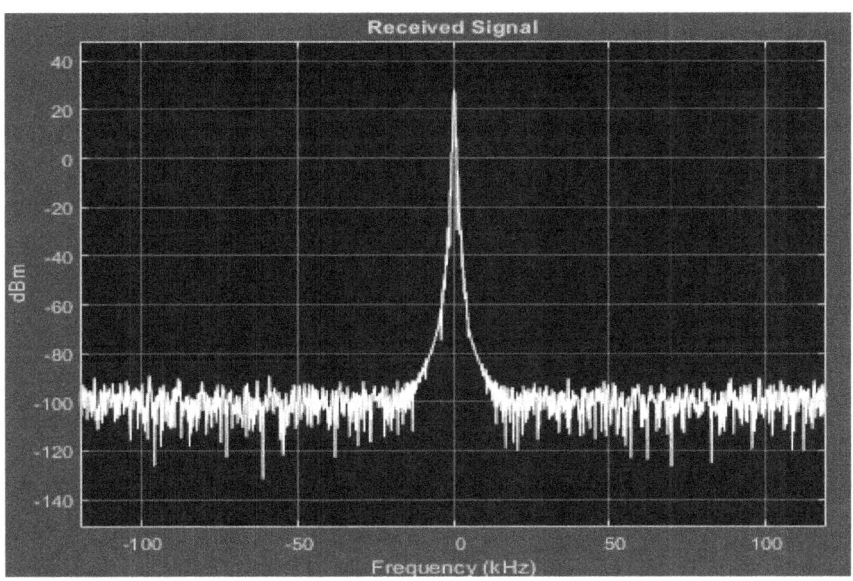

Having grasped the basic functionalities of your GMRS radio and understood the performance-influencing factors, it's time to truly master the language of frequencies, signals, and protocols. This chapter equips you with the knowledge to confidently navigate GMRS radio operations.

We'll begin by exploring the invisible highways your voice travels on – the operating frequencies and channels that form the foundation of communication. You'll learn how to transmit and receive signals effectively, ensuring your messages are delivered clearly and efficiently. Understanding proper channel etiquette and protocols is crucial for responsible and respectful communication within the GMRS community.

Next, we'll examine the technical aspects of power output and modulation, the building blocks that determine the strength and clarity of your transmissions. This chapter will introduce you to the world of advanced modulation schemes, including Frequency Modulation (FM) and Phase Modulation (PM), providing a deeper understanding of how your GMRS radio translates your voice into transmittable signals.

The journey doesn't end there. Understanding the behavior of radio waves and the potential for interference is essential for optimizing your communication strategies. This chapter will equip you with the knowledge to navigate potential challenges and ensure your voice reaches its destination loud and clear.

By the end of this chapter, you'll be well-prepared to operate your radio confidently, navigate channels, and enjoy clear, reliable communication on your next adventure.

3.1 Operating Frequencies and Channels: Demystifying the Invisible Infrastructure of GMRS Communication

Imagine a bustling marketplace filled with vendors hawking their wares. Each vendor occupies a designated stall to avoid chaos and ensure clear communication with their customers. Similarly, GMRS radio communication thrives on a well-organized system of frequencies and channels, acting as the invisible infrastructure that keeps your conversations clear and efficient. Let's embark on a journey to understand these fundamental concepts.

3.1.1 GMRS Frequency Band: Your Designated Marketplace

Just like vendors need a physical space to operate, GMRS radios require a specific portion of the radio spectrum to function. The Federal Communications Commission (FCC) allocates designated frequency bands for various communication purposes. Think of these frequency bands as designated marketplaces within the vast electromagnetic spectrum. GMRS radios operate within a specific segment of the Very High Frequency (VHF) band, typically ranging from 462 MHz to 467 MHz. Imagine this band as a bustling marketplace specifically for GMRS radio communication. Radios within this range can "hear" each other, allowing for conversations to take place.

3.1.2 Channels: Your Designated Stall Within the Marketplace

While the GMRS frequency band provides the overall marketplace for communication, it would be quite chaotic if everyone tried to talk simultaneously. To ensure order and prevent conversations from interfering with each other, the GMRS band is further divided into 22 narrow channels, each spaced slightly apart. Imagine these channels as designated stalls within the larger GMRS marketplace. By setting your radio to a specific channel, you're essentially choosing your stall. Everyone else using the same channel is essentially at the same stall, allowing you to communicate with them clearly. Think of it like tuning into a specific radio station – you only hear the program broadcasted on that frequency, not the programs on other stations.

3.1.3 Selecting the Right Channel: Etiquette for a Harmonious Marketplace

Unlike some other radio services with assigned channels, GMRS operates on a shared-use basis. This means any GMRS radio user can utilize any of the available 22 channels. However, to ensure smooth communication and avoid the cacophony of overlapping conversations, a community approach is essential. Here are some key points to remember:

- **Channel Monitoring: Be a Courteous Customer** – Imagine approaching a busy stall in the marketplace – you wouldn't want to barge in and interrupt an ongoing conversation. Similarly, before transmitting on a channel, listen for ongoing conversations using the "monitor" function on your radio. If the channel is clear, that's your cue to proceed with your message.
- **Respectful Communication: Sharing the Space** – Just like any marketplace thrives on respectful interactions, so too does GMRS communication. Avoid lengthy conversations or monopolizing a channel. Be mindful of others who might need to use the channel and keep your transmissions concise and informative.
- **Local Agreements: Navigating Unwritten Marketplace Rules** – In some areas, GMRS users might have established informal agreements regarding channel usage. For example, a specific channel might be designated for local events or off-road communication by enthusiasts in your area. Think of it as unwritten rules within the local marketplace. Checking with local GMRS users or online forums can help you discover any such agreements in your area, ensuring you operate within the established norms.

Understanding operating frequencies, channels, and proper channel etiquette, puts you well on your way to becoming a responsible and respectful member of the GMRS radio community. The next section will dig deeper into the process of transmitting and receiving signals, exploring how your voice travels through the air and reaches its destination.

3.2 Transmitting and Receiving Signals: The Magic Behind the Message

Now that you understand the invisible highways (frequencies) and designated lanes (channels) of GMRS radio communication, let's explore the fascinating process of transmitting and receiving signals. This section will unveil the magic behind how your voice travels through the air and reaches its destination.

3.2.1 From Voice to Radio Waves: The Power of Modulation

When you speak into your GMRS radio, the microphone picks up the sound waves generated by your voice. These sound waves are analog signals, meaning they continuously vary in amplitude (volume) and frequency (pitch) to represent your speech. However, radio waves cannot directly transmit these analog signals. Here's where modulation comes in – a process that transforms your analog voice signal into a radio wave suitable for transmission over the air.

Imagine a carrier wave – a basic radio wave with a constant frequency and amplitude. During transmission, the information from your voice signal (amplitude and frequency variations) is impressed upon this carrier wave using a modulation technique. Think of it like adding texture and color to a blank canvas – the carrier wave becomes the canvas, and your voice signal adds the details. In GMRS radios, a specific type of modulation called Narrowband FM (NFM) is commonly used.

3.2.2 Unveiling the Magic of NFM: Riding the Waves of Your Voice

Narrowband FM, or NFM, works by varying the frequency of the carrier wave in accordance with the amplitude variations of your voice signal. Imagine the pitch (frequency) of the carrier wave rising and falling as you speak louder or softer. This creates a new radio wave that carries the information of your voice within its frequency variations.

Here's an analogy: Think of riding a bicycle – when you pedal harder, you go faster (increased amplitude), which can be likened to an increase in the carrier wave's frequency. Conversely, pedaling slower (decreased amplitude) translates to a decrease in the carrier wave's frequency. By encoding your voice information this way, the NFM-modulated radio wave can travel efficiently through the air.

3.2.3 Receiving the Signal: Decoding the Message

When an NFM-modulated radio wave reaches the receiving GMRS radio, the process is reversed. The receiver uses a demodulator to extract the original information from the carrier wave. The demodulator essentially separates the voice information (encoded frequency variations) from the carrier wave, returning it to its original analog form. This recovered analog signal is then amplified and converted back into audible sound waves by the speaker of the receiving radio, allowing you to hear the transmitted message clearly.

3.2.4 Putting It All Together: The Cycle of Communication

Imagine you're talking to a friend over a walkie-talkie, but instead of a simple on/off switch, there's a whole process happening behind the scenes to get your voice from your mouth to their ear. Here's a breakdown of that process:

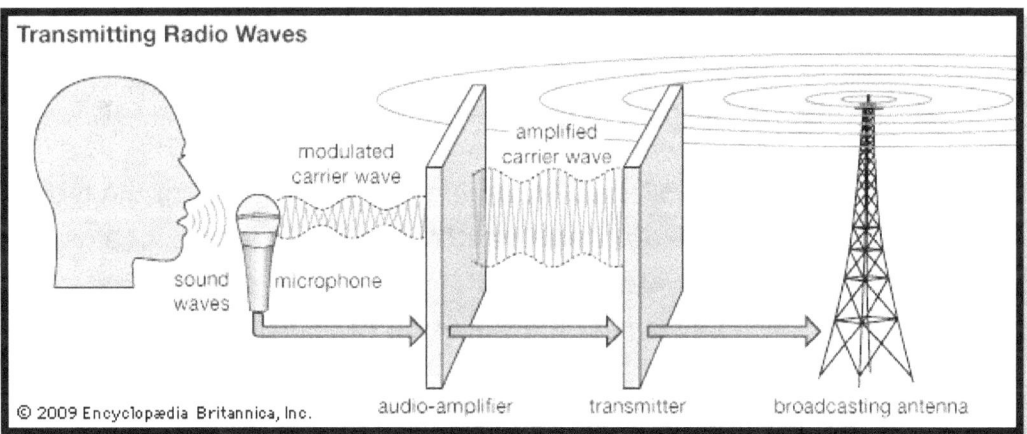

1. **Your Voice Makes a Signal:** When you speak into the microphone, your voice creates sound waves. These waves are like wiggles in the air, with the height of the wiggle showing how loud you're talking (amplitude) and the space between wiggles showing how high-pitched your voice is (frequency).

2. **Shaping Your Voice for the Journey (Modulation):** Radios can't directly send those sound waves. Instead, your GMRS radio uses a technique called Narrowband FM (NFM) to put your voice on a special radio wave called a carrier wave. Think of the carrier wave like a flat, featureless highway, and your voice signal like a bumpy

road. NFM takes the ups and downs of your voice signal (amplitude and frequency) and uses them to slightly bend the carrier wave up and down in a specific way. This puts the information about your voice onto the carrier wave.

3. **Sending Your Voice on a Radio Wave (Transmission):** Once your voice is on the carrier wave, it's ready for the trip! Your radio boosts the power of the signal and sends it out through the antenna. The antenna acts like a launcher, pushing the radio wave out into the air.

4. **Catching the Signal (Reception):** The radio of the person you're talking to has an antenna too. This antenna acts like a receiver, picking up the radio wave carrying your voice information.

5. **Unwrapping Your Voice (Demodulation):** The radio then uses a demodulator, which is like a special decoder ring. The demodulator reads the ups and downs of the carrier wave and translates them back into the ups and downs of your original voice signal (amplitude and frequency).

6. **Hearing Your Voice (Back to Analog):** Finally, the radio strengthens the signal again and sends it to a speaker, which turns the electrical signal back into sound waves. This lets the other person hear your voice just like you spoke it!

Understanding this process supplies you with a deeper appreciation for the technology that powers your GMRS radio and facilitates clear communication across the airwaves. The next section will dig into the crucial aspects of proper channel etiquette and protocols, ensuring you operate your radio responsibly within the GMRS community.

3.3 Proper Channel Etiquette and Protocols: Navigating the Shared Highway of GMRS Communication

Imagine a multi-lane highway bustling with traffic. While designated lanes ensure order, courteous drivers and established traffic rules are essential for a smooth and safe journey. Similarly, navigating the shared channels of GMRS radio communication requires proper etiquette and adherence to protocols to ensure clear communication and a positive experience for everyone.

3.3.1 Respectful Communication: The Cornerstone of Etiquette

- **Listen Before You Transmit:** Just like checking for oncoming traffic before changing lanes, always listen attentively to the chosen channel before transmitting. Avoid interrupting ongoing conversations and ensure a clear break in communication before proceeding with your message.
- **Keep it Concise:** Remember, GMRS channels are a shared resource. Strive for clear and concise transmissions, avoiding lengthy conversations or unnecessary chatter that could clog the channel for others.
- **Identify Yourself:** It's good practice to identify yourself briefly at the beginning and end of your transmissions. This helps others on the channel know who they are communicating with and adds a layer of professionalism.
- **Mind Your Language:** Maintain a professional and courteous tone during communication. Avoid offensive language or topics that could be inappropriate for a public forum.

3.3.2 Understanding Call Signs and Over: Essential Tools for Communication

- **Call Signs:** While not mandatory for all GMRS users, having a Federal Communications Commission (FCC) issued call sign adds a layer of formality and allows for easier identification during communication. Think of it as your designated license plate on the GMRS highway.
- *Over.* This brief term is used to indicate your intention to transmit. Saying "Over" at the end of your message signals to others that you have finished speaking and relinquishing the channel for potential responses. Imagine it as a turn signal, letting others know when it's safe to "merge" into the conversation.

3.3.3 Regional Agreements and Common Courtesy:

In some areas, local GMRS users might have established informal agreements regarding channel usage. For example, a specific channel might be designated for off-road communication or local events. Checking with local GMRS communities or online forums can help you discover any such agreements in your area. Following these unwritten rules demonstrates respect for the local community and ensures smooth communication.

By adhering to these etiquette guidelines and protocols, you contribute to a positive and respectful environment on the GMRS radio channels. The next section will explore the technical aspects of power output and modulation, exploring the factors that influence the strength and clarity of your transmissions.

3.4 Power Output and Modulation: Optimizing Your Voice for the Journey

Now that you've mastered the language of channel etiquette and protocols, let's look into the technical aspects of your GMRS radio. This section will explore two key factors that influence the reach and clarity of your transmissions: power output and modulation.

3.4.1 Power Output: How Much Muscle Does Your Radio Have?

Imagine shouting across a vast canyon. The strength of your voice determines how far it travels. Similarly, GMRS radios have a specific power output rating, measured in watts, that indicates the strength of the radio waves they can transmit. Higher power output generally translates to a greater communication range, allowing your voice to travel farther and reach stations at a greater distance.

However, there are limitations to consider. The FCC regulations governing GMRS radios restrict the maximum power output to 5 watts. Additionally, factors like terrain and obstacles can significantly impact the actual range you experience, even with a higher power output radio.

3.4.2 Modulation: Shaping Your Voice for the Journey

As we learned earlier (Section 3.2), modulation plays a crucial role in transforming your voice signal into a radio wave suitable for transmission. In GMRS radios, a specific type of modulation called Narrowband FM (NFM) is used. NFM works by varying the frequency of the carrier wave in accordance with the amplitude variations of your voice.

While NFM is a robust and reliable modulation technique, it's important to understand that it doesn't inherently boost the power of your transmission. The power output of your radio determines the overall strength of the radio wave, and NFM shapes the information your voice carries onto that wave.

3.4.3 Finding the Right Balance: Power Output and Communication Strategies

Since GMRS radios have a limited power output, maximizing your communication range often involves strategies beyond simply cranking up the power. Here are some key considerations:

- **Understanding Terrain:** Hills, mountains, and buildings can significantly weaken your signal. If possible, try to communicate from higher ground with a clearer line of sight to the receiving station.
- **Proper Antenna Usage:** A well-maintained antenna with a higher gain can help focus your radio wave in a specific direction, potentially increasing its effective range.
- **Respecting Regulations:** Remember, GMRS regulations limit your power output to 5 watts. Operating outside these limits is illegal and can lead to hefty fines.

By understanding the interplay between power output and modulation, you can develop effective communication strategies to ensure your voice reaches its destination clearly, even within the limitations of GMRS radio regulations. The next section will introduce you to the world of advanced modulation schemes, offering a glimpse into potential future advancements in GMRS communication.

3.5 Advanced Modulation Schemes: A Look Beyond Narrowband FM

While Narrowband FM (NFM) is the workhorse modulation technique for GMRS radios, the world of radio communication offers a variety of more advanced modulation schemes. This section will provide a brief glimpse into these advancements and their potential implications for future GMRS technology.

3.5.1 Frequency Modulation (FM): A Broader Perspective

NFM is a specific type of Frequency Modulation (FM). In FM, the frequency of the carrier wave is varied to represent the information being transmitted. However, NFM utilizes a relatively narrow range of frequency deviation, prioritizing clarity over potential range benefits.

More advanced FM schemes, like Wideband FM (WBFN), employ a wider range of frequency deviation. This can potentially offer some advantages, such as increased resistance to interference and improved audio quality. However, WBFN is not currently permitted for use in GMRS radios due to its broader signal footprint, which could lead to increased channel congestion.

3.5.2 Phase Modulation (PM): A Different Approach to Encoding Information

Phase Modulation (PM) is another modulation technique that can be used to transmit information. In PM, the information signal causes variations in the phase of the carrier wave, rather than its frequency. PM offers advantages like improved spectral efficiency and potential resistance to certain types of noise.

However, PM can be more susceptible to certain other types of interference compared to FM. Currently, PM is not used in GMRS radios, and its potential future adoption would depend on advancements in technology and regulatory considerations.

3.5.3 The Future of GMRS Modulation: Balancing Needs and Regulations

The current regulations governing GMRS radios prioritize reliable, clear communication within a limited power output and spectrum allocation. NFM strikes a good balance between these factors. While advanced modulation schemes like WBFN or PM offer potential benefits, their broader signal footprints or increased complexity might not be suitable for the shared-use environment of GMRS channels.

The future of GMRS modulation might involve continued refinement of NFM or the development of new, hybrid techniques that maintain the core strengths of NFM while incorporating advancements for improved performance within the existing regulatory framework.

3.5.4 The Importance of Understanding the Basics

While this section provided a glimpse into advanced modulation schemes, it's important to remember that a strong foundation in NFM is crucial for effective GMRS radio communication. Understanding the interplay between power output and modulation techniques empowers you to optimize your communication strategies within the capabilities of your GMRS radio.

The next section will explore the fascinating world of radio wave behavior and the potential for interference. This understanding will equip you to navigate challenges and ensure your voice reaches its destination loud and clear.

3.6 Understanding Radio Wave Behavior and Interference: Navigating the Invisible Landscape

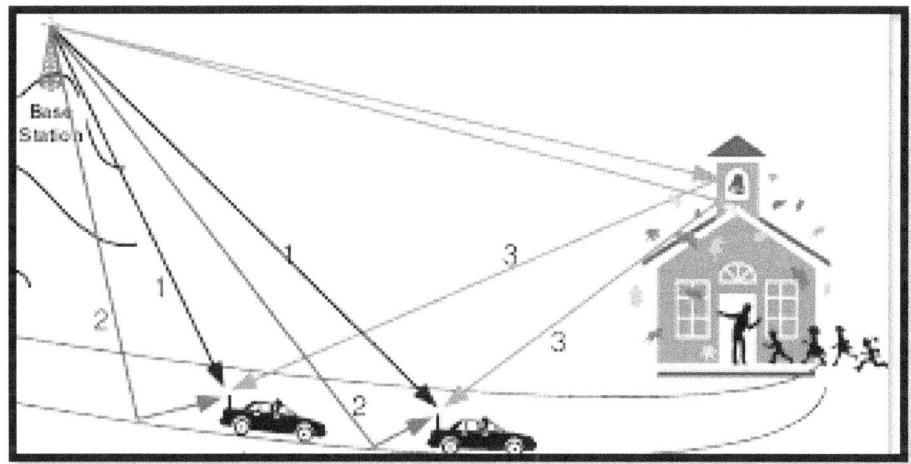

We've embarked on a fascinating journey, exploring the inner workings of GMRS radios, proper communication etiquette, and the role of power output and modulation. Now, it's time to explore the invisible world of radio waves themselves. Understanding how radio waves behave and the potential for interference is essential for optimizing your communication strategies and ensuring your message reaches its destination clearly.

3.6.1 Radio Waves on the Move: Not Quite Like Light

Unlike light waves that travel in a straight line, radio waves bend and diffract around obstacles. Imagine throwing a pebble into a pond – ripples emanate outwards in all directions. Radio waves behave similarly, following multiple paths as they travel from the transmitting antenna to the receiving antenna.

There are three main modes of radio wave propagation that influence how your GMRS signal travels:

- **Line-of-Sight Propagation:** In ideal scenarios with a clear line of sight between the transmitter and receiver, radio waves travel directly, offering the most efficient mode of propagation and the maximum communication range. Think of an unobstructed view between two points – this is the ideal scenario for clear and strong signals.
- **Ground Wave Propagation:** Radio waves can also travel along the surface of the earth, hugging the ground like a wave following the coastline. This propagation mode is particularly effective for lower frequencies, including those used in GMRS radios. However, the ground acts as a conductor, absorbing some of the signal energy, leading to a gradual weakening of the signal strength with increasing distance. Imagine radio waves encountering resistance as they travel along the ground, causing the signal to diminish over long distances.
- **Skywave Propagation:** At higher frequencies, radio waves can reflect off layers of the ionosphere, a region of the upper atmosphere charged with electrically charged particles. Imagine a radio wave bouncing off a giant mirror in the sky. This reflection allows for communication over vast distances, but it's a less reliable mode of propagation for GMRS radios due to the unpredictable nature of the ionosphere. Think of a constantly changing reflective surface – skywave propagation can be unreliable for GMRS communication.

3.6.2 The Challenge of Interference: Sharing the Airwaves

Just like multiple cars navigating a busy highway, radio waves can interfere with each other. Here are two main types of interference that can potentially disrupt your GMRS communication:

- **Signal Collision:** Imagine two cars trying to occupy the same lane on the highway – chaos ensues! Similarly, if two strong radio signals are on the same frequency within your reception range, they can collide and disrupt the clarity of your received message.
- **Noise:** Electrical devices, power lines, and even natural phenomena like lightning can generate electrical noise that disrupts radio signals. Think of loud, distracting sounds on the highway – noise can make it difficult to hear clear communication on your GMRS radio.

3.6.3 Overcoming the Challenges: Strategies for Clear Communication

Understanding radio wave behavior and the potential for interference empowers you to develop strategies for clear communication:

- **Utilize Higher Ground:** If possible, communicate from higher points like hills or rooftops. This provides a clearer line of sight for the radio waves and can significantly improve communication range, especially in areas with challenging terrain.
- **Respect the Limitations:** GMRS radios have a limited range. Don't rely solely on them for critical situations where long-distance or absolutely reliable communication is paramount. Consider alternative communication methods as backups in case of signal limitations.
- **Minimize Interference:** Turn off unnecessary electronics near your radio that might generate noise. Be aware of potential sources of electrical interference in your environment.

Summary

This chapter transformed you from a basic GMRS radio user to someone who understands the language and can navigate the invisible landscape of radio wave communication. We started by exploring the concept of frequencies and channels, the invisible highways that carry your voice. You learned how to select the right channel and the importance of proper channel etiquette for respectful and efficient communication.

Next, we ventured into the technical aspects of transmitting and receiving signals. You unveiled the magic of modulation, how your voice is transformed into a radio wave for transmission, and how it's decoded back into sound at the receiving end. Understanding power output and its limitations further equipped you to develop effective communication strategies.

The chapter then introduced you to the world of advanced modulation schemes, offering a glimpse into potential future advancements in GMRS technology. While NFM remains the workhorse for GMRS radios, understanding its role and the concept of modulation empowers you to optimize your communication even within the limitations of your radio's capabilities.

The journey didn't stop there. We ventured into the fascinating world of radio wave behavior, exploring how radio waves travel and the potential for interference from signal collision and noise. By understanding these invisible forces, you can develop strategies like utilizing higher ground or minimizing interference to ensure your voice reaches its destination loud and clear.

By equipping you with this knowledge, this chapter has empowered you to confidently operate your GMRS radio, navigate channels effectively, and optimize your communication strategies for clear and reliable experiences, wherever your adventures take you. The next chapter will focus on the legalities and regulations governing GMRS radio use, ensuring you operate responsibly within the set guidelines.

Review Questions

1. Describe two key aspects of proper channel etiquette that contribute to a smooth and respectful communication experience for all users.
2. While both power output and modulation influence the strength and clarity of your GMRS radio transmissions, they serve different purposes. Explain the role of power output and how modulation, specifically NFM, shapes the information carried on the radio wave.
3. Radio waves travel differently than sound waves, and their behavior can impact your GMRS communication range. Describe two ways understanding radio wave propagation, such as line-of-sight and ground wave propagation, can help you develop strategies for clear communication.

CHAPTER 4

GMRS RADIO SETUP AND INSTALLATION

Now that you've mastered the language of GMRS communication and understand the fascinating world of radio waves, it's time to put your knowledge into action! This chapter digs into the exciting world of setting up and installing your GMRS radio. Whether you're incorporating a mobile unit into your vehicle or opting for a handheld option, this chapter equips you with the essential steps for a successful installation.

We'll begin by exploring the proper techniques for installing and mounting mobile units in your vehicle, ensuring optimal placement and functionality. Next, we'll tackle the crucial task of programming channels and frequencies, allowing you to access the specific channels designated for GMRS communication in your area.

As a vital component for clear communication, antenna selection and optimization will be thoroughly discussed. You'll learn how to choose the right antenna for your needs and how to position it for the best possible signal reception and transmission.

Understanding how to power your GMRS radio is essential, and this chapter will guide you through various power source options and effective battery management strategies. For those seeking advanced features, we'll explore the integration of GPS technology and the potential benefits of location services offered by some GMRS radios.

Finally, the chapter will equip you with strategies to mitigate RF (Radio Frequency) interference, a potential challenge in some environments. By understanding the sources of interference and implementing effective countermeasures, you can ensure clear and uninterrupted communication on your GMRS radio.

Throughout this chapter, we'll provide clear instructions, helpful tips, and best practices to ensure your GMRS radio setup is not only successful but also optimized for your specific needs and communication goals. So, grab your radio, get ready to unleash its full potential, and let's embark on this journey of setting up your reliable communication companion!

4.1 Installing and Mounting Mobile Units: Finding the Perfect Spot for Optimal Performance

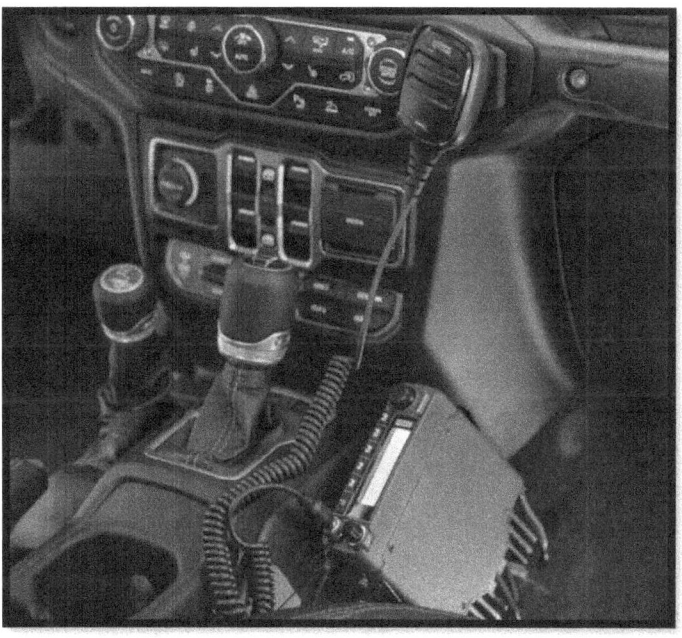

Equipping your vehicle with a GMRS radio transforms it into a mobile communication hub. This section will guide you through the process of installing and mounting your mobile unit, ensuring it's secure, functional, and conveniently accessible.

Understanding Your Mobile Unit:

Before diving into installation, take a moment to familiarize yourself with your specific GMRS radio model. Identify the connection points for the antenna, power cable, microphone (if applicable), and any external speaker connections. Consult the user manual for your radio to understand the recommended mounting options and any specific installation requirements.

Planning for Placement:

The key to successful mobile unit installation lies in strategic placement. Here are some crucial factors to consider:

- **Accessibility:** The radio should be within easy reach for safe and convenient operation while driving. Avoid locations that might obstruct your view of the road or hinder your ability to control the vehicle. Consider your dominant hand and typical driving posture when choosing a mounting location.
- **Visibility:** The radio display should be clearly visible, allowing you to monitor channel information, signal strength, and other relevant data without straining your eyes. This is important for maintaining focus on the road while staying informed about your GMRS communication.
- **Cable Management:** Think about the path your cables will take from the radio to the antenna, power source, and microphone (if applicable). Plan for a clean and organized cable layout to prevent them from becoming tangled or obstructing movement within the vehicle.
- **Mounting Options:** Many mobile units come with mounting brackets that allow for secure attachment to various surfaces within your vehicle. Depending on your preference and the available space, you can choose from dash mounts, console mounts, or even headrest mounts. Some aftermarket mounting solutions offer additional features like adjustability or swivel functionality for increased ergonomic comfort.

The Installation Process:

Once you've identified the ideal location, follow these general steps for installation:

1. **Prepare the Mounting Surface:** Clean and dry the chosen mounting area to ensure a strong and secure bond. Refer to the user manual or mounting bracket instructions for any specific requirements regarding surface preparation.
2. **Mount the Radio Unit:** Secure the radio unit to the mounting bracket according to the manufacturer's instructions. This might involve screws, clips, or a combination of both, depending on the design.
3. **Connect the Cables:** Route the antenna cable, power cable, and microphone cable (if applicable) neatly to their respective connection points on the radio unit. Double-check the connections to ensure they are secure and properly aligned.
4. **Organize the Cables:** Use zip ties or cable management solutions to keep the cables organized and prevent them from dangling or interfering with other components within the vehicle.
5. **Test the Unit:** Power on your GMRS radio and verify that all functions are working correctly. Make sure the display is visible, buttons respond properly, and the microphone transmits your voice clearly (if applicable). Test the antenna connection by transmitting a test message to another GMRS radio to confirm functionality.

Additional Considerations:

- **Power Source:** We'll discuss power source options in detail in Section 4.4, but during installation, consider the proximity of your chosen power source to the mounting location. This will help you determine the appropriate cable length needed.
- **Professional Installation:** While installing a mobile GMRS radio is typically a straightforward process, if you're uncomfortable with any aspect of the installation or require assistance with running cables through specific areas of your vehicle, consider consulting a professional car audio installer.

Follow these guidelines and carefully plan your mobile unit installation to ensure your GMRS radio is securely mounted, conveniently accessible, and ready to connect you to your communication network on the go. The next section will discuss the crucial task of programming your radio for optimal functionality.

4.2 Programming Channels and Frequencies: Demystifying the Control Panel of Your GMRS Radio

Imagine your GMRS radio as a high-tech walkie-talkie, but instead of a single channel, it has access to a designated range of frequencies for GMRS communication. This section will guide you through the process of programming your radio to access these specific channels and frequencies, ensuring you can connect with others using the GMRS network.

Understanding Channels and Frequencies:

Think of channels as virtual "lanes" on a highway dedicated for GMRS communication. Each channel is assigned a specific frequency, which is the actual radio wave used to transmit and receive information. Your GMRS radio allows you to select a particular channel, and the radio automatically tunes to the corresponding frequency for that channel.

Locating Channel Information:

The FCC (Federal Communications Commission) regulates the use of GMRS frequencies in the United States. There are a total of 22 designated GMRS channels, each with a corresponding frequency. To program your radio, you'll need to know the specific channel numbers and frequencies used in your area. Here are some resources to help you find this information:

- **FCC GMRS website:** The FCC website provides a comprehensive overview of GMRS regulations, including a table listing all 22 GMRS channels and their corresponding frequencies.
- **Local GMRS Users Group:** Connecting with a local GMRS users group in your area can be a valuable resource. These groups often maintain information about commonly used channels in your specific location. You can find local GMRS users groups online through forums, social media groups, or by contacting your local ham radio clubs.

Programming Your Radio:

The specific steps for programming your radio will vary depending on the model you own. Most GMRS radios come with a user manual that provides detailed instructions on navigating the programming menu and setting channels. Here's a general overview of the process:

1. **Power On Your Radio:** Turn on your GMRS radio and locate the programming menu button(s). Consult your user manual for specific button combinations or menu options related to programming.
2. **Access Channel Programming:** Once you've located the programming menu, navigate to the section dedicated to channel programming. This might be labeled as "Channels," "Frequencies," or a similar term depending on your radio model.
3. **Enter Channel Number:** The programming menu will likely prompt you to enter a channel number. Use the keypad or control buttons on your radio to select the desired channel number from the available range (typically 1 to 22 for GMRS radios).
4. **Enter Frequency (if applicable):** Some radios might require you to enter the specific frequency corresponding to the channel. If this is the case, consult your channel information and use the keypad or control buttons to enter the correct frequency for the chosen channel.
5. **Save and Repeat:** Once you've entered the channel information, save the settings according to your radio's user manual instructions. Repeat this process for each channel you want to program on your radio.

Additional Tips:

- **Consult the User Manual:** The user manual for your specific GMRS radio model is your best resource for detailed programming instructions. Refer to the manual for any specific steps, menu options, or button combinations unique to your radio.
- **Start with a Few Channels:** For beginners, it's helpful to start by programming just a few commonly used channels in your area. This simplifies operation and allows you to become familiar with the programming process before adding more channels.

- **Channel Labeling:** Some radios allow you to assign custom labels to programmed channels. This can be a helpful way to identify frequently used channels or channels dedicated to specific communication groups (e.g., "Family Channel," "Off-Road Group").

- **Practice Makes Perfect:** Programming your radio might take a few tries at first. Don't hesitate to consult your user manual or online resources if you encounter any difficulties.

Follow these steps and familiarize yourself with the programming process and you'll be well on your way to unlocking the communication potential of your GMRS radio. The next section probes the crucial role of your antenna and how to optimize it for clear and reliable communication.

4.3 Antenna Installation and Optimization: Tuning in for Crystal-Clear Communication

Your GMRS radio is the communication hub, but the antenna acts as its voice and ears. A well-chosen and properly installed antenna is essential for transmitting and receiving clear signals. This section will guide you through selecting the right antenna for your needs and optimizing its placement for maximum performance.

Understanding Antenna Basics:

Antennas work by converting electrical signals from your radio into radio waves for transmission and vice versa. The size, design, and placement of the antenna all influence its effectiveness in transmitting and receiving radio waves. Here are some key factors to consider when choosing a GMRS antenna:

- **Gain:** Antenna gain refers to its ability to strengthen the signal in a particular direction. Higher gain antennas are generally better for long-range communication, while lower gain antennas might be suitable for shorter-range communication in areas with less obstruction.
- **Length:** The length of the antenna is typically related to its operating frequency. For GMRS radios operating in the UHF band (around 462-467 MHz), antennas tend to be shorter compared to antennas for lower frequencies. However, keep in mind that a longer antenna within a reasonable range for GMRS use will generally offer better performance.
- **Type:** There are various types of antennas available for GMRS radios, each with its own advantages and disadvantages. Here's a brief overview of some common types:
  - **J-Pole Antennas:** These are versatile antennas known for their omnidirectional radiation pattern, meaning they transmit and receive signals equally in all directions. They are a popular choice for handheld radios and mobile units due to their compact size and ease of use.
  - **Vertical Antennas:** These antennas offer a more focused radiation pattern, with better signal strength in the vertical direction (up and down). They can be a good choice for mobile units mounted on vehicles, especially if you plan to communicate with others at similar elevations.
  - **Directional Antennas:** These antennas offer the most focused signal in a specific direction. They are ideal for long-range communication in situations where you know the location of the other party you want to

communicate with. However, they require precise aiming for optimal performance.

Matching Your Needs:

The best antenna for your GMRS radio depends on your specific needs and communication goals. Consider the following factors when making your selection:

- **Portability vs. Range:** If portability is a priority (e.g., using a handheld radio for hiking), a J-Pole antenna might be a good choice. If long-range communication is your main focus, a higher gain vertical antenna or a directional antenna could be better options.
- **Mounting Location:** The location where you plan to mount the antenna will influence your choice. J-Pole antennas are versatile for both handheld and mobile use. For mobile units mounted on vehicles, consider the clearance from the roof or other parts of the vehicle when choosing an antenna length.
- **Local Regulations:** In some areas, there might be regulations regarding antenna height or type for GMRS use. Check with your local authorities or GMRS user groups to ensure your chosen antenna complies with any local restrictions.

Installation and Optimization:

Once you've selected your antenna, follow these steps for proper installation:

1. **Refer to the Manual:** Consult the user manual for both your GMRS radio and your chosen antenna for specific installation instructions. These instructions will typically detail the mounting process and any necessary adjustments.
2. **Mount the Antenna:** Securely mount the antenna in the chosen location according to the manufacturer's instructions. Pay close attention to proper grounding requirements, if applicable, to ensure optimal performance.
3. **Positioning (if applicable):** For antennas with directional radiation patterns, adjust the positioning based on the direction of your expected communication. Consult the antenna's user manual for guidance on achieving the desired signal focus.

4. **Test and Fine-Tune:** If possible, test your antenna setup with another GMRS radio user to assess signal strength and reception quality. Some GMRS radios might have built-in signal strength meters that can provide a basic indication of antenna performance. You can make minor adjustments to the antenna position (if applicable) to fine-tune performance.

Additional Considerations:

- **Ground Plane:** The performance of some antenna types, particularly mobile antennas mounted on vehicles, can be affected by the presence of a good ground plane. A ground plane is a large metal surface, such as the roof of a car, that can improve antenna efficiency. Consult your antenna's user manual for recommendations regarding ground plane requirements.

Coaxial Cable: The coaxial cable connects your antenna to your GMRS radio. Choose a cable with the appropriate length and impedance (typically 50 ohms for GMRS applications) to minimize signal loss during transmission and reception. Higher quality coaxial cables with lower signal loss ratings will offer better performance, especially for longer cable runs.

- **Replacement Antennas:** Most GMRS radios come with a basic antenna included. Upgrading to a higher gain or more specialized antenna can significantly improve your communication range and reception quality.

By understanding antenna basics, carefully selecting the right antenna for your needs, and following proper installation and optimization techniques, you can ensure your GMRS radio has the best possible "ears" for clear and reliable communication. The next section will address the crucial aspect of powering your GMRS radio and explore strategies for effective battery management.

4.4 Power Source and Battery Management: Keeping Your GMRS Radio Juiced Up

Your GMRS radio is a powerful communication tool, but just like any electronic device, it needs a reliable source of power to function. This section will guide you through understanding your power options and developing effective strategies for managing your GMRS radio's battery life.

Understanding Power Sources:

There are two main ways to power your GMRS radio:

- **Batteries:** This is the most common power source for handheld GMRS radios. Most radios utilize rechargeable batteries, typically Nickel-Metal Hydride (NiMH) or Lithium-Ion (Li-Ion) batteries. These batteries can be recharged multiple times before needing replacement. Some handheld radios might also have the option of using disposable alkaline batteries, but this is generally not recommended due to higher running costs and potential environmental impact.
- **External Power Source:** Mobile GMRS units designed for vehicle use typically come with a DC power cable that allows you to connect the radio directly to your vehicle's electrical system. This provides a constant power source for extended use while on the go. Some handheld radios might also offer the option of using an external power bank or a similar portable power source for extended operation in the field.

Choosing the Right Battery:

When it comes to batteries for your handheld GMRS radio, consider these factors:

- **Battery Capacity:** Measured in milliamp-hours (mAh), battery capacity indicates how long the battery can power your radio on a single charge. Higher mAh ratings generally translate to longer operating times. Choose a battery capacity that aligns with your typical usage patterns.
- **Battery Type:** As mentioned earlier, NiMH and Li-Ion are the most common rechargeable battery types for GMRS radios. NiMH batteries are generally more affordable but might offer slightly shorter lifespans and require more frequent charging cycles compared to Li-Ion batteries. Li-Ion batteries typically offer higher capacity, longer lifespans, and faster charging times, but they might come at a slightly higher cost.

- **Replacement Batteries:** Consider having a spare battery on hand, especially if you anticipate extended use away from a charging source. This ensures you can keep your communication flowing without interruption.

Battery Management Tips:

Extend the life of your GMRS radio batteries and ensure reliable operation by implementing these practices:

- **Proper Charging:** Always follow the manufacturer's instructions for charging your GMRS radio batteries. Avoid overcharging, which can damage the battery's lifespan. Unplug the charger once the battery is fully charged.
- **Minimize Backlight Usage:** The display backlight on your GMRS radio can consume a significant amount of battery power. Reduce backlight brightness or turn it off completely when not actively using the radio to conserve battery life.
- **Turn Off the Radio When Not in Use:** Don't leave your GMRS radio powered on when you're not actively using it. Develop the habit of turning it off to maximize battery life, especially between transmissions.
- **Temperature Extremes:** Extreme temperatures, both hot and cold, can negatively impact battery performance. Store your GMRS radio and spare batteries in moderate temperatures whenever possible. If using your radio in cold weather conditions, consider carrying a spare battery stored closer to your body heat to maintain its performance.
- **Monitor Battery Level:** Most GMRS radios have a battery level indicator. Pay attention to this indicator and recharge your battery before it becomes completely depleted. Letting the battery completely drain can reduce its overall lifespan.

External Power Considerations:

If you plan to use an external power source for your mobile GMRS radio, keep these points in mind:

- **DC Power Adapter Compatibility:** Ensure the DC power adapter you use is compatible with your specific GMRS radio model and has the correct voltage output. Using an incompatible adapter can damage your radio.
- **Cable Length:** Choose a DC power cable with sufficient length to reach your vehicle's power outlet comfortably without creating excessive strain on the connection points.
- **Power Management:** While an external power source provides continuous operation, it's still a good practice to power down your radio when not in use to conserve battery life and reduce wear on the electrical components.

Understanding your power options, choosing the right batteries, and implementing effective battery management strategiess ensures that your GMRS radio is always ready to connect you when you need it most. The next section will explore the exciting world of GPS integration and the potential benefits of location services offered by some GMRS radios.

4.5 GPS Integration and Location Services (if applicable)

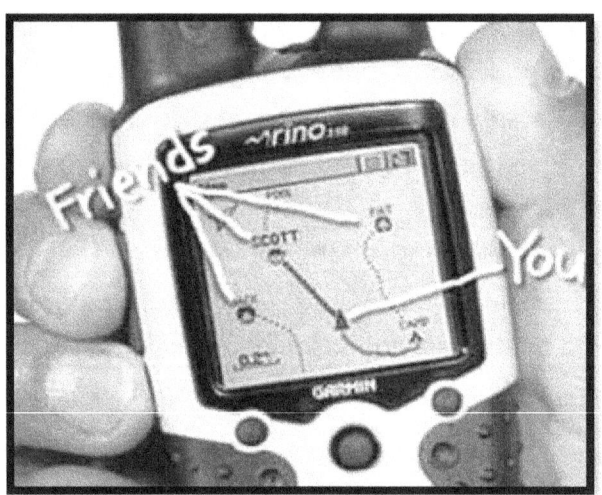

Not all GMRS radios come equipped with GPS functionality. However, if your radio boasts this feature, it can add a valuable layer of safety and convenience to your communication. This section will guide you through setting up and utilizing the GPS integration on your GMRS radio.

Before you Begin:

- **Consult your Radio Manual:** The specific steps for enabling and using GPS features will vary depending on your radio model. Always refer to the manufacturer's instructions for the most accurate information.
- **Identify GPS Compatibility:** Check your radio's user manual or product information to confirm it has built-in GPS. Look for keywords like "GPS integration," "location sharing," or "geolocation."

Enabling GPS and Location Services:

1. **Power On Your Radio:** Turn on your GMRS radio according to the manufacturer's instructions.
2. **Access Settings Menu:** Locate the button or menu option for accessing your radio's settings. This might be labeled "Settings," "Menu," or a dedicated button with an icon.
3. **Navigate to GPS Settings:** Within the settings menu, look for options related to GPS, location, or geolocation. The specific wording may vary depending on your radio model.
4. **Enable GPS:** Once you locate the relevant menu, find the option to enable or activate GPS functionality. This might be a simple on/off toggle or require additional configuration.
5. **Set Location Sharing Preferences (Optional):** Some radios may allow you to configure location sharing with other compatible devices. Review the options and choose your preference based on privacy and safety considerations.

Using GPS Features:

- **Basic Location Information:** Depending on your radio, you might be able to view your current coordinates (latitude and longitude) on the display. This can be helpful for pinpointing your location on a map or relaying it to others.
- **Location Sharing:** If your radio supports location sharing, you might be able to transmit your location data to other compatible radios within a certain range. This

can be a valuable safety feature when coordinating group activities or in emergency situations.

- **Location-based Applications (Advanced):** Some high-end GMRS radios with GPS integration might offer additional features like mapping software or integration with smartphone apps. These advanced applications would require referring to the specific model's user manual for proper setup and use.

Important Considerations:

- **GPS Accuracy:** While GPS is generally reliable, accuracy can be affected by factors like signal strength, terrain, and urban environments. Be aware of these limitations when relying on GPS data for critical situations.
- **Battery Consumption:** Using GPS functionality can drain your radio's battery life more quickly. Be mindful of this and manage your battery usage accordingly, especially on extended outings.
- **Privacy Settings:** Review and adjust location sharing settings based on your comfort level and the specific situation. It's important to be mindful of who has access to your location data.

By following these steps and considering the important points, you can leverage the benefits of GPS integration on your GMRS radio for enhanced communication and safety during your adventures.

4.6 RF Interference Mitigation Strategies

Even in remote locations, your GMRS radio can encounter Radio Frequency (RF) interference. This interference can manifest as static, crackling, or even complete loss of signal. Here, we'll explore some strategies to mitigate RF interference and ensure clear communication on your GMRS radio.

Identifying Interference:

- **Listen for Signs:** The first step is recognizing the presence of interference. This might include static, crackling noises, or intermittent signal drops during conversations.

- **Identify the Source (if possible):** Sometimes, the source of interference can be obvious. This could be nearby high-voltage power lines, electrical equipment, or even other GMRS radios operating on the same channel.

Minimizing Interference:

- **Adjust Antenna Position:** A well-positioned antenna can significantly improve signal reception and reduce interference. Try adjusting the antenna's angle or height for optimal performance. Refer to your radio's manual for proper antenna placement recommendations.
- **Change Channels:** GMRS radios operate on a designated set of channels. If you're experiencing interference on your current channel, try switching to a different one. You might find clearer communication on another frequency.
- **Increase Distance:** If the source of interference is identifiable (e.g., another radio user), try increasing the distance between your radios to reduce signal overlap.
- **Utilize Repeaters (if available):** GMRS repeater stations can amplify your signal and extend your communication range. If a repeater is within range and compatible with your radio, using it can help bypass localized interference.

Other Considerations:

- **Weather Conditions:** Dense fog, heavy rain, or even solar flares can affect radio signal propagation. Be aware of potential weather impacts on communication clarity.
- **Radio Quality:** Using a high-quality GMRS radio with a strong receiver can help mitigate interference compared to lower-end models.

Note: Complete elimination of RF interference might not always be possible. However, by employing these strategies, you can significantly improve your chances of achieving clear and reliable communication on your GMRS radio.

Summary

Chapter 4: GMRS Radio Setup and Installation provides a comprehensive guide for users to get their GMRS radios up and running for optimal performance.

- **Mobile Unit Installation:** The chapter starts by outlining the process of installing and securely mounting your GMRS radio in a vehicle for hands-free operation while on the go.
- **Channel Programming:** It then discusses how to program your radio with the appropriate channels and frequencies for clear communication within the designated GMRS spectrum.
- **Antenna Optimization:** Proper antenna installation and positioning are crucial for receiving strong signals and maximizing your radio's range. The chapter offers guidance on achieving optimal antenna setup.
- **Power Source Management:** Understanding your radio's power requirements and implementing efficient battery management strategies are covered to ensure reliable communication during extended use.
- **Optional GPS Integration:** For radios equipped with GPS functionality, the chapter details how to enable and utilize location services for enhanced safety and potential location sharing features.
- **Mitigating Interference:** The chapter concludes by acknowledging the potential for Radio Frequency (RF) interference and offers strategies to minimize its impact, ensuring clear and uninterrupted communication on your GMRS radio.

Review Questions

1. The chapter discussed programming channels and frequencies on your GMRS radio. Can you explain the importance of using the correct channel and the potential consequences of using an incorrect one?
2. The guide mentioned antenna optimization for better signal reception. Describe two factors to consider when positioning your GMRS radio antenna for maximum range and performance.
3. For radios with GPS functionality, the chapter offered instructions on enabling location services. Can you identify two potential benefits of using GPS features on your GMRS radio during communication?

Chapter 5: GMRS Radio Communication Techniques

Having your GMRS radio set up and ready to go is only half the battle. To truly unlock its potential, you need to master the art of clear, effective communication. This chapter dives deep (no pun intended!) into the essential techniques and strategies that will transform you from a novice user to a confident GMRS communicator.

We'll begin by exploring the fundamentals of establishing clear communication channels, ensuring everyone hears you loud and clear. Next, we'll explore the world of call signs and identifiers, the essential tools for identifying yourself and others on the network.

For those seeking an extra layer of security, we'll explore advanced encryption and privacy features available on some GMRS radios. This section will equip you with the knowledge to protect sensitive conversations.

The chapter then expands its reach by exploring the possibilities of remote operation and control. We'll discuss both remote head systems, allowing you to operate your radio from a distance, and smartphone integration, enabling convenient control through your mobile device.

Finally, for those venturing into situations where clear and coordinated communication is paramount, we'll unveil tactical communication strategies and protocols. By learning these techniques, you'll be prepared to handle emergencies or group activities with confidence and efficiency.

So, buckle up and get ready to elevate your GMRS communication skills to the next level! This chapter will equip you with the knowledge and tools to navigate the world of GMRS radio communication with clarity, security, and effectiveness.

5.1 Establishing Clear Communications

The foundation of any successful GMRS radio communication lies in establishing clear and concise connections. This section digs into the essential practices that ensure your message is received loud and clear, avoiding misunderstandings and wasted transmissions.

Pre-Transmission Checks:

- **Power On and Volume Check:** Before initiating any communication, ensure your radio is powered on and the volume is adjusted to a comfortable level for both sending and receiving messages.
- **Channel Selection:** Verify you're on the appropriate GMRS channel designated for your intended communication. Reviewing channel charts or programming presets on your radio can help avoid confusion.
- **Antenna Check:** A properly functioning antenna is crucial for clear transmission and reception. Ensure your antenna is securely mounted and undamaged.

Initiating Communication:

- **Station Identification:** Start your transmission by announcing your call sign or radio ID clearly. This allows others on the channel to identify you.
- **Attention Signal (Optional):** Some radios offer an attention signal option, like a short tone burst. This can be used to grab the attention of others on the channel before your message. However, use this sparingly to avoid disrupting ongoing conversations.
- **State Your Intent:** Clearly state your purpose for transmitting. Are you trying to establish contact with a specific station or group? Are you relaying information relevant to everyone on the channel?

Transmission Practices:

- **Clarity and Concision:** Speak clearly and concisely at a moderate pace. Enunciate your words properly to avoid misunderstandings.
- **Proper Use of Pauses:** Brief pauses between sentences and phrases can enhance clarity, especially in noisy environments.
- **Minimize Background Noise:** Try to transmit from a location with minimal background noise like engine sounds or wind. Move to a quieter spot if necessary.

Receiving Transmissions:

- **Active Listening:** Pay close attention to incoming transmissions. When receiving a call, allow the speaker to finish their message before responding.
- **Acknowledge Receipt:** Once you've received a message, briefly acknowledge receipt to indicate you understand and are ready to respond or continue the conversation.

Maintaining Communication Flow:

- **Take Turns Talking:** Effective communication is a two-way street. Avoid talking over others and allow for proper turn-taking during conversations.
- **Minimize Background Transmissions:** Avoid unnecessary background conversations or side-chats while others are using the channel.
- **Keep it Relevant:** Focus your transmissions on relevant information related to the purpose of your communication.

Following these guidelines enables you to establish a strong foundation for clear and effective GMRS radio communication. Remember, courtesy and consideration for other users go a long way in maintaining a positive communication environment on the shared GMRS channels.

5.2 Using Call Signs and Identifiers

In the world of GMRS radio communication, just like CB radios or amateur radio, clear identification is crucial. This section explores the concept of call signs and identifiers, the tools that allow everyone on the channel to know who they're talking to.

Understanding Call Signs:

- **Unique Identifier:** A GMRS call sign is a unique alphanumeric code assigned to a licensed GMRS user or station. It serves as your official identification on the network.

- **Obtaining a Call Sign:** To acquire a GMRS call sign, you need to obtain a Federal Communications Commission (FCC) license. The FCC application process is typically straightforward and can be completed online.
- **Proper Use of Call Signs:** When initiating communication, clearly state your call sign at the beginning of your transmission. You should also include the call sign of the station you're trying to contact if known.

Alternative Identifiers:

- **Radio ID:** Some GMRS radios allow you to program a shorter Radio ID for convenience. This ID can be used instead of your full call sign during casual communication, as long as everyone on the channel is familiar with it.
- **Group IDs:** For group communication, a designated Group ID can be used to identify a specific group or team operating on the channel. This helps differentiate group conversations from general broadcasts.

Benefits of Using Identifiers:

- **Reduced Confusion:** Proper use of call signs and identifiers eliminates confusion on the channel, ensuring everyone knows who is transmitting and who they are addressing.
- **Improved Safety:** Clear identification is particularly important in emergency situations, allowing responders to quickly identify those in need of assistance.
- **Network Etiquette:** Using identifiers demonstrates proper radio etiquette and promotes a more professional communication environment.

Additional Considerations:

- **Privacy Concerns:** While call signs are a matter of public record, some users might prefer to use Radio IDs for casual conversations to maintain a level of privacy.
- **International Considerations:** If traveling near international borders, be aware that GMRS call signs might not be recognized in neighboring countries. It's important to research local regulations for short-range radio communication.

By understanding and properly utilizing call signs and identifiers, you'll contribute to a more organized and efficient GMRS radio communication network. Do not forget that clear identification is not just a formality, it's a safety measure and a courtesy to your fellow GMRS users.

5.3 Advanced Encryption and Privacy Features

While GMRS radio communication generally operates on open channels, some advanced models offer features for enhanced security and privacy. This section explores these features and their potential benefits for specific communication needs.

Understanding Encryption:

- **Scrambling Messages:** Encryption technology scrambles your voice transmission into an unreadable format. Only authorized users with the appropriate decryption key can understand the message content. This adds a layer of security for sensitive conversations.
- **Types of Encryption:** There are various encryption algorithms used in GMRS radios. Some models might offer basic analog scrambling, while others might utilize more sophisticated digital encryption methods.

Privacy Features:

- **Digital Codes:** Certain radios allow you to program privacy codes. These codes act like digital filters that block unwanted transmissions on the same channel. Only radios programmed with the same privacy code will be able to decode and hear your conversation.
- **Voice Activation (VOX):** Voice-activated transmission (VOX) allows you to transmit hands-free by automatically activating the microphone only when you speak. This can be a privacy benefit as it eliminates the risk of accidentally broadcasting background conversations or noises.

Benefits of Encryption and Privacy Features:

- **Protecting Sensitive Information:** Encryption is crucial for safeguarding confidential communication, especially for commercial use or situations where sensitive information might be exchanged.
- **Enhanced Privacy:** Privacy features like digital codes can offer a layer of privacy during casual conversations, reducing the chance of unintended listeners receiving your message.
- **Reduced Interference:** Privacy codes can help minimize unwanted background noise or overlapping transmissions from other users on the same channel.

Important Considerations:

- **Limited Availability:** Encryption and advanced privacy features are not available on all GMRS radios. These features are typically found on higher-end models.
- **Legal Restrictions:** In some regions, there might be legal restrictions on the use of encryption technology in civilian radio communication. It's important to be aware of local regulations before utilizing encryption features.
- **Compatibility Issues:** Encryption and privacy features only work between radios programmed with the same settings. Ensure all parties involved in a secure conversation have compatible radios with matching encryption algorithms or privacy codes.

Making the Choice:

The decision to utilize encryption or privacy features depends on your specific communication needs. If you require maximum security for sensitive conversations, encryption is the way to go. However, for casual communication where a basic level of privacy is desired, digital codes or VOX might suffice. Remember, even with these features, GMRS radio communication still operates on shared channels, so complete privacy cannot be guaranteed.

5.4 Remote Operation and Control

The world of GMRS radios extends beyond traditional handheld units mounted in vehicles. This section explores two advanced options for remote operation and control, offering greater flexibility and convenience in your communication experience.

5.4.1 Remote Head Systems

- **Separate Control Unit:** A remote head system consists of two separate components: a control unit and a remotely mounted transceiver unit. The control unit, typically a compact and user-friendly handheld device, houses the microphone, speaker, display, and controls for operating the radio.
- **Remote Mounting:** The transceiver unit, containing the transmitter and receiver hardware, is mounted remotely, often near the antenna for optimal signal strength. This separation allows for more flexible placement, particularly beneficial in vehicles where space might be limited.

Benefits of Remote Head Systems:

- **Improved Ergonomics:** The handheld control unit provides a comfortable and convenient interface for operating the radio, especially when mounted within easy reach for hands-free communication.
- **Flexible Placement:** Remote mounting of the transceiver unit allows for better antenna placement, potentially improving signal quality. This is advantageous in situations where a roof-mounted antenna might be impractical.
- **Enhanced Safety:** In some vehicles, a remote head system allows for safer radio operation by keeping the control unit within easy reach, eliminating the need to reach for a mounted radio while driving.

Considerations for Remote Head Systems:

- **Cost:** Remote head systems typically cost more than traditional handheld radios.
- **Installation:** Proper installation of the transceiver unit and cabling might require professional assistance.

- **Compatibility:** Ensure the remote head system is compatible with your specific GMRS radio model.

5.4.2 Smartphone Integration

TEXT MESSAGE

The rise of smartphones has opened doors for innovative ways to interact with GMRS radios. This section explores the concept of smartphone integration and its potential benefits.

- **Bluetooth Connectivity:** Certain GMRS radios offer Bluetooth connectivity, allowing them to pair with your smartphone. This creates a powerful communication hub.
- **Dedicated Apps:** Manufacturers often develop dedicated mobile applications for their GMRS radios. These apps provide a user-friendly interface for controlling the radio remotely from your smartphone.

- **Expanded Features:** Smartphone apps can unlock a range of functionalities beyond basic radio controls. Features might include real-time location sharing, text messaging over GMRS (where available), or access to online call logs and channel information.

Benefits of Smartphone Integration:

- **Convenience and Control:** Smartphone integration allows you to control your GMRS radio remotely, offering greater flexibility and ease of use.
- **Enhanced Features:** Mobile apps can unlock additional functionalities, expanding the capabilities of your GMRS radio.
- **Improved Information Management:** Apps can provide features like call logs and channel information, streamlining your communication experience.

Considerations for Smartphone Integration:

- **Compatibility:** Ensure your GMRS radio and smartphone are compatible for Bluetooth pairing and app functionality.
- **Data Usage:** Using a smartphone app for GMRS communication might incur data charges depending on your mobile plan.
- **Battery Consumption:** Both the radio and smartphone battery life might be impacted by using Bluetooth connectivity and running communication apps.

Choosing the Right Option:

The choice between a remote head system and smartphone integration depends on your specific needs and preferences. If you prioritize a dedicated control unit and flexible antenna placement, a remote head system might be ideal. However, if convenience, expanded features, and smartphone integration are your focus, a Bluetooth-enabled GMRS radio with a dedicated app could be the better choice.

5.5 Tactical Communication Strategies and Protocols

Effective communication is paramount in situations requiring coordinated action and clear information flow. This section dives into tactical communication strategies and protocols specifically designed to enhance GMRS radio usage in critical scenarios.

Planning and Preparation:

- **Pre-defined Channels:** Establish designated channels for communication within your group before venturing into a potentially challenging situation. This ensures everyone is on the same page and avoids confusion during crucial moments.
- **Clear and Concise Messages:** Adhere to the principles of clear and concise communication as outlined in Section 5.1. In high-pressure situations, brevity and clarity are essential for avoiding misunderstandings.
- **Standardized Terminology:** Develop and agree upon a set of common phrases and abbreviations for frequently used terms. This streamlines communication and reduces the risk of misinterpretations.
- **Communication Hierarchy:** If your group is large or complex, establish a communication hierarchy with designated leaders or spokespersons responsible for relaying information. This prevents a cacophony of voices and ensures clear direction.

Communication Protocols:

- **Call Before Transmit (CBT):** Implement a Call Before Transmit (CBT) protocol. This requires users to announce their call sign or identifier before transmitting any message, even brief ones. This helps maintain order and avoid interruptions during critical moments.
- **Roger/Wilco:** Utilize standardized acknowledgment codes like "Roger" (message received and understood) or "Wilco" (message received and will comply) to confirm message reception and intent.
- **Periodic Check-ins:** Depending on the situation, establish a schedule for periodic check-ins with all team members. This ensures everyone is accounted for and facilitates information exchange.

- **Emergency Protocols:** Define clear protocols for emergency situations. This might involve designated emergency channels, pre-arranged distress calls, and established procedures for reporting emergencies and coordinating response efforts.

Maintaining Situational Awareness:

- **Minimize Non-essential Traffic:** Avoid transmitting unnecessary chatter that can clog the channel and hinder critical communication. Focus on relaying essential information relevant to the situation at hand.
- **Monitor the Channel:** Team members should actively monitor the designated channel to stay updated on the situation and respond to instructions or requests for assistance.
- **Adaptability:** Be prepared to adapt communication strategies based on the evolving situation. Flexibility is key to ensuring effective communication in dynamic environments.

Implementing these tactical communication strategies and protocols helps you significantly enhance the effectiveness of your GMRS radio usage during critical situations. Remember, clear and coordinated communication can be a vital element in ensuring safety, success, and a positive outcome in challenging scenarios.

Summary

Chapter 5: **GMRS Radio Communication Techniques,** equips you to transform your GMRS radio from a simple device into a powerful communication tool.

This chapter dives deep into essential techniques and strategies to elevate you from a novice user to a confident GMRS communicator. Key takeaways include:

- **Establishing Clear Communication:** Learn proper practices for ensuring clear transmissions, minimizing misunderstandings, and maintaining a positive communication environment.
- **Using Call Signs and Identifiers:** Grasp the importance of call signs and identifiers for clear identification on the channel and proper radio etiquette.

- **Advanced Features: Encryption & Privacy (Optional):** Explore encryption and privacy features available on some radios for enhanced security in sensitive communication scenarios.
- **Remote Operation and Control:** Discover the flexibility and convenience offered by remote head systems and smartphone integration for GMRS radio control.
- **Tactical Communication Strategies:** Master the art of clear and coordinated communication in critical situations with established protocols and best practices for emergencies and group activities.

By mastering the valuable concepts presented in this chapter, you'll be well-equipped to leverage the full potential of your GMRS radio and communicate effectively in any situation, from casual conversations to demanding adventures.

Review Questions

1. The chapter discussed the "Call Before Transmit (CBT)" protocol. Explain the importance of CBT and how it contributes to clear and efficient communication, especially in group settings.
2. The importance of clear and concise communication was emphasized throughout the chapter. Describe two additional strategies mentioned in the chapter, besides using concise messages, that can further enhance clarity during GMRS radio communication.
3. For users concerned about privacy, the chapter explored encryption and privacy features. However, these features might not be available on all radios. Identify two alternative strategies users can implement to add a layer of privacy to their GMRS radio communication, even without encryption.

CHAPTER 6

ADVANCED FEATURES AND FUNCTIONS

Having mastered the fundamentals of GMRS radio communication in the previous chapter, we now embark on a journey to explore the exciting realm of advanced features and functions. These capabilities elevate your GMRS radio beyond basic communication, transforming it into a feature-rich tool for enhanced functionality and versatility.

This chapter will meticulously examine a range of advanced functionalities, equipping you to unlock the full potential of your GMRS radio. We'll look at:

- **Scanning Modes and Features:** Discover how to efficiently search for active channels and conversations using various scanning methods.
- **Dual Watch and Priority Scan:** Learn to monitor two channels simultaneously or prioritize a specific channel while scanning for activity on others.
- **CTCSS and DCS Codes:** Explore these subaudible tone technologies that offer a layer of privacy and reduce channel clutter.
- **Weather Channels and Alerts:** Understand how to utilize weather channels and receive critical weather information updates on your GMRS radio.
- **Multi-User and Group Communication:** Dive into advanced features for facilitating communication within groups, including group call and private call functionalities.
- **Data Transmission and Packet Radio Techniques:** (For technically-inclined users) Explore the possibilities of data transmission and packet radio techniques for more advanced communication needs.

By the conclusion of this chapter, you'll possess a comprehensive understanding of the advanced features embedded within your GMRS radio, enabling you to leverage its full capabilities and tailor your communication experience to your specific needs. So, buckle up and prepare to be amazed by the hidden potential of your GMRS radio!

6.1 Scanning Modes and Features

Not all conversations happen on the same channel at the same time. Scanning modes empower you to efficiently search for active channels and conversations on the GMRS frequency band. This section will unveil the various scanning functionalities offered by most GMRS radios.

- **Basic Scanning:** This is the most straightforward mode. The radio systematically scans through all programmed channels, pausing briefly on each one to check for activity. If a transmission is detected, the radio might stop on that channel for a user-defined duration or until the transmission ends.
- **Priority Channel Scanning:** This mode prioritizes a specific channel while still scanning others. You can designate a frequently used channel or a channel reserved for emergencies as your priority. The radio will check the priority channel at regular intervals, even while scanning other channels. If activity is detected on the priority channel, the scan will pause, and the transmission will be audible. Once the transmission ends, the radio will resume scanning other channels.
- **Programmed Scan:** This mode allows you to scan only a specific subset of pre-programmed channels within the GMRS band. This is useful if you only use a limited number of channels regularly and want to avoid scanning through unused frequencies.
- **Tone Squelch Scan:** This advanced scanning mode combines scanning with squelch functionality (explained in Chapter 4). The radio scans for active channels, but it will only unmute and become audible if a specific CTCSS or DCS tone (covered in Section 6.3) is detected alongside the transmission. This helps eliminate unwanted noise from inactive channels and focuses your attention on relevant conversations using your preferred privacy codes.
- **Search and Lock:** This mode allows you to manually search for active channels outside of your programmed list. By manually adjusting the frequency, you can scan for transmissions beyond the standard GMRS channels. However, it's important to be aware of regulations regarding communication outside of licensed frequencies.

Understanding these scanning modes empowers you to efficiently search for active conversations, stay connected with your group, and avoid missing important messages on the GMRS network. Remember to consult your specific GMRS radio's user manual for detailed instructions on navigating its scanning functionalities.

6.2 Dual Watch and Priority Scan

Building upon the foundation of scanning modes, this section explores two advanced functionalities designed to enhance your ability to monitor multiple channels simultaneously: Dual Watch and Priority Scan.

- **Dual Watch:**

Imagine having two conversations happening at once! Dual Watch allows you to monitor two pre-selected channels concurrently. Your radio will typically have a designated button or menu option to activate Dual Watch mode. Once enabled, the radio will seamlessly switch back and forth between your chosen channels at a user-defined scan rate. This allows you to stay updated on activity on both channels without manually switching back and forth.

Here are some scenarios where Dual Watch can be beneficial:

- **Monitoring a Group Channel and a Calling Channel:** You can keep an ear on your group's designated channel for ongoing communication while also monitoring a separate calling channel for potential incoming calls from outside your group.
- **Tracking Two Locations:** If you have team members operating in two different areas, Dual Watch allows you to monitor activity on both channels, ensuring you don't miss updates from either location.
- **Priority Scan:**

While Dual Watch monitors two channels equally, Priority Scan offers a more nuanced approach. This mode assigns a designated "priority channel" that takes precedence over other programmed channels during scanning. The radio will continuously scan through all programmed channels except for the priority channel. However, if activity is detected on the priority channel, the scan will pause, and the transmission will be audible. Once the transmission ends, the scan will resume, checking all other channels except for the priority channel again.

Here are some situations where Priority Scan can be useful:

- **Monitoring a Calling Channel with Active Channels:** You can set a frequently used calling channel or an emergency channel as your priority. This ensures you don't miss important incoming calls while still scanning for activity on other channels.
- **Staying Alert for Urgent Communication:** In situations where you anticipate critical updates, assigning the designated emergency channel as your priority ensures you'll be immediately alerted if a transmission occurs on that channel.

Strategically utilizing Dual Watch and Priority Scan significantly enhances your ability to stay informed and manage communication across multiple channels on your GMRS radio. Remember to consult your radio's user manual for specific instructions on activating and configuring these features.

6.3 CTCSS and DCS Codes: Decoding the Language of Subaudible Tones

Imagine a world where you can hold a conversation on a busy street corner without being overheard by everyone around you. In the realm of GMRS radio communication, CTCSS and DCS codes achieve a similar feat, creating a layer of privacy and control over who hears your transmissions and who you hear on a channel. This section dives deep into the fascinating world of these subaudible tones, equipping you with a comprehensive understanding of their inner workings.

The Realm of the Inaudible:

The human ear is a marvel of biological engineering, but it has limitations. We can typically detect sound waves within a frequency range of 20 Hz to 20,000 Hz. Frequencies below 20 Hz are perceived as infrasound, while those exceeding 20,000 Hz fall into the realm of ultrasound, both inaudible to the human ear. CTCSS (Continuous Tone-Coded Squelch System) and DCS (Digital Coded Squelch) exploit this very principle.

- **CTCSS: The Analog Maestro:** CTCSS utilizes a set of continuous low-level tones, typically ranging from 67 Hz to 254.1 Hz, which are embedded with your voice transmission. These tones are so faint that they fall outside the audible range for humans, existing silently alongside your voice message.
- **DCS: The Digital Architect:** DCS (Digital Coded Squelch) takes a different approach. Instead of continuous tones, it employs digital codes – brief bursts of data – modulated at a carrier frequency of around 134 Hz. Similar to CTCSS, this carrier frequency itself is inaudible, but it carries the embedded digital code, acting as the secret handshake for your communication.

The Magic Behind the Mute Button: Squelch Explained

Every GMRS radio has a built-in squelch circuit. When the squelch is closed (off), your radio's speaker remains silent, even if there's activity on the channel. This prevents your radio from constantly emitting static or background noise when no usable signal is present.

The beauty of CTCSS and DCS lies in their interaction with the squelch. Your GMRS radio can be programmed with a specific CTCSS tone or DCS code. When you transmit, the chosen subaudible tone or digital code is invisibly interwoven with your voice signal. The radio's receiver also has a decoder function programmed to listen for a specific CTCSS tone or DCS code.

- **The CTCSS Handshake:** If the received signal carries a continuous tone that perfectly matches the programmed CTCSS tone in your radio, it's like finding the right key. The decoder recognizes the match and triggers the squelch to open. Voila! The audio becomes audible, allowing you to hear the incoming transmission. If the received signal doesn't have the matching CTCSS tone, the decoder remains silent, and the squelch stays closed, effectively muting any unwanted noise from other conversations on the same channel.

- **The DCS Digital Check:** For DCS, the process is equally ingenious. The decoder in your radio deciphers the embedded digital code within the received signal. If the code precisely matches the one you programmed, it's like passing a digital security check. The decoder throws the switch, and the squelch opens, allowing the audio to flow through. Any transmissions with DCS codes that don't match your programmed code are treated like imposters by the decoder, and the squelch remains firmly closed, blocking out the unwanted noise.

The Benefits of Subaudible Tones: A Multi-faceted Advantage

By incorporating CTCSS or DCS codes into your GMRS radio communication strategy, you unlock several advantages:

- **Reduced Channel Clutter:** Imagine a busy highway filled with cars. CTCSS and DCS codes act like lane filters for your radio. With these codes enabled, your radio only becomes active (squelch opens) for transmissions carrying your designated code, effectively filtering out unwanted conversations happening on the same channel. This translates to less background chatter and a clearer listening experience.
- **Enhanced Privacy:** While not foolproof, CTCSS and DCS codes offer a basic layer of privacy for casual conversations on shared GMRS channels. Only users programmed with the same CTCSS tone or DCS code will be able to hear your transmissions, creating a sense of exclusivity within the channel. It's like having a private conversation in a crowded room by using a pre-arranged code word that only your intended recipient understands.
- **Improved Organization for Groups:** For groups utilizing GMRS radios, CTCSS or DCS codes can be a valuable tool for organizing communication. By assigning a specific code to your group, you can ensure that transmissions within your group are clear and uninterrupted by other conversations on the same channel. This streamlines communication and minimizes confusion.

Choosing the Right Code: CTCSS vs. DCS

Both CTCSS and DCS offer similar benefits, but there are some key factors to consider when choosing which system to use:

*   **Number of Codes:** CTCSS offers a limited set of analog tones (typically around 38-50 codes depending on the radio model). DCS, on the other hand, utilizes digital codes, providing a significantly larger pool of possibilities (often exceeding 100 codes). This wider range of codes in DCS can be advantageous in areas with high GMRS radio usage, reducing the chance of accidentally using the same code as another group.
*   **Interference Resistance:** Analog CTCSS tones can sometimes be susceptible to interference from electrical noise or bleed-over from strong signals on nearby frequencies. DCS, with its digital codes, offers better resistance to such interference, ensuring more reliable decoding and clearer communication.
*   **Radio Compatibility:** Ensure your GMRS radios are compatible with the chosen system (CTCSS or DCS) and offer the specific number of tones or codes you require. Not all radios support both systems, and some might have limitations on the available code options.
*   **Local Regulations:** In some regions, there might be regulations regarding the use of CTCSS or DCS codes. It's advisable to consult local regulations or licensing authorities to ensure your chosen codes are compliant.

Important Considerations: Understanding the Limitations

While CTCSS and DCS codes offer advantages, it's crucial to understand their limitations:

- **Not Foolproof Security:** These codes do not provide absolute privacy. Anyone with a GMRS radio scanner can detect activity on a channel regardless of the code used. For truly secure communication, consider encryption technologies (if available on your radio model) or alternative communication methods that offer built-in encryption.
- **Code Selection Matters:** Choosing a commonly used code might defeat the purpose of privacy, as other users on the channel might be programmed with the same code. Select a less common code for enhanced privacy, especially in areas with high GMRS radio activity.
- **Interference Can Disrupt Decoding:** Even DCS, with its better resistance, can be susceptible to extreme interference in rare cases. This could lead to missed transmissions if the decoder fails to recognize the digital code due to signal distortion.

By understanding the intricacies of CTCSS and DCS codes, their functionalities, limitations, and the factors influencing code selection, you can leverage these subaudible tones to enhance your GMRS radio communication experience. Remember, these codes are a valuable tool for reducing channel clutter, improving clarity, and adding a basic layer of privacy to your conversations, but they should not be considered a failsafe for secure communication.

6.4 Weather Channels and Alerts: Harnessing the Power of Real-Time Weather Information

The saying goes, "knowledge is power," and nowhere is this more true than when it comes to weather. For outdoor enthusiasts and anyone who relies on being prepared for changing weather conditions, GMRS radios equipped with weather channels and alert capabilities can be a game-changer. This section will navigate you through the world of weather information accessible on your GMRS radio.

Understanding NOAA Weather Radio (NWR):

Most GMRS radios come equipped with a built-in receiver for the National Oceanic and Atmospheric Administration (NOAA) Weather Radio (NWR) network. NWR is a nationwide network of radio stations that continuously broadcast weather information directly from the nearest National Weather Service office. This real-time information can include:

- **Current weather conditions:** Get instant updates on temperature, humidity, wind speed and direction, precipitation, and other relevant weather parameters for your local area.
- **Forecasts:** Stay informed about upcoming weather patterns with short-term and extended forecasts, allowing you to plan your activities accordingly.
- **Hazardous weather alerts:** NWR broadcasts critical weather alerts, such as warnings for tornadoes, severe thunderstorms, flash floods, blizzards, and other imminent weather hazards. These timely alerts can be lifesaving, giving you the opportunity to take precautions and seek shelter if necessary.

Utilizing Weather Channels on Your GMRS Radio:

- **Finding the Weather Channel:** Most GMRS radios with NWR reception functionality have a dedicated weather channel button or menu option. Consult your radio's user manual for specific instructions on locating and activating the weather channel.
- **Weather Alert Tones and Indicators:** Many radios feature built-in alert tones or visual indicators that activate when a hazardous weather alert is broadcasted. This ensures you don't miss critical weather warnings even if the radio is muted.
- **Programming Specific Alert Types:** Some advanced radios allow you to program specific types of weather alerts you wish to receive. For example, you might choose to be notified only for tornado warnings or flash flood alerts relevant to your location.

The Benefits of Weather Information on GMRS Radios:

- **Enhanced Situational Awareness:** Real-time weather updates and forecasts empower you to make informed decisions about your outdoor activities. You can plan hikes, camping trips, or other adventures with confidence, knowing the weather conditions you can expect.
- **Improved Safety:** Weather alerts provide crucial warnings about potential hazards, allowing you to take necessary precautions and seek shelter during severe weather events. This can significantly reduce the risk of accidents and injuries.
- **Peace of Mind:** Having access to up-to-date weather information directly on your GMRS radio fosters peace of mind, especially for those venturing into remote locations. The ability to stay informed about changing weather patterns adds a layer of security to your outdoor activities.

Important Considerations:

- **NWR Coverage:** While NWR boasts extensive coverage across the United States, there might be areas with limited or no signal reception. It's advisable to check NWR coverage maps for your specific location.
- **Backup Plans:** Don't solely rely on NWR for weather information. Always carry a battery-powered weather radio or a portable NOAA weather alert receiver as a backup, especially in areas with potential NWR signal limitations.
- **Understanding Alert Messages:** Weather alerts can contain technical jargon. Familiarize yourself with common weather alert terminology to ensure you can interpret the warnings accurately and take appropriate action.

Effectively utilizing the weather channels and alert functionalities of your GMRS radio helps you harness the power of real-time weather information. This knowledge empowers you to make informed decisions, prioritize safety, and enjoy your outdoor adventures with greater peace of mind.

6.5 Multi-User and Group Communication: Expanding the Reach of Your GMRS Radio

The beauty of GMRS radio communication lies in its ability to connect you with others. This section looks into advanced features designed to facilitate communication within groups, extending the reach and functionality of your GMRS radio beyond one-on-one conversations.

6.5.1 Group Call Features:

Imagine coordinating a group hike or managing a team during an outdoor event. Group call features on your GMRS radio empower you to communicate with multiple individuals simultaneously, streamlining coordination and information flow within your group. Here's how it works:

- **Pre-programmed Group ID:** Each member of your group programs their GMRS radio with a designated group ID. This ID acts like a digital channel specifically for your group.
- **Initiating a Group Call:** A designated member or anyone within the group can initiate a group call by selecting the group ID and transmitting their message.
- **Receiving Group Calls:** All GMRS radios programmed with the matching group ID will receive the transmission, allowing everyone in the group to hear the message simultaneously.

Benefits of Group Call Features:

- **Enhanced Coordination:** Group calls facilitate efficient communication within your team. Information dissemination, issuing instructions, and receiving updates from multiple members become effortless.
- **Streamlined Communication:** No need to call individual members one by one. Group calls allow for quick updates and discussions with the entire group, saving time and effort.
- **Improved Situational Awareness:** Everyone in the group receives the same information simultaneously, ensuring all members are on the same page and can adapt to changing situations as a cohesive unit.

6.5.2 Private Call Features:

While group calls are ideal for broadcasting messages to the entire team, situations might arise where private communication between two members is necessary. Here's where private call features come into play:

- **Digital Selective Calling (DSC):** This functionality allows you to send a digital call signal containing a specific identification code to another GMRS radio programmed with the same DSC capability. Upon receiving the DSC, the designated recipient's radio will typically emit a distinct tone or alert, notifying them of a private call request.
- **Direct Calling with Call Aliases:** Some advanced GMRS radios offer direct calling functionalities with pre-assigned call aliases (nicknames) for identified users within your group. This allows for point-to-point communication with a specific member without interrupting the entire group conversation.

Benefits of Private Call Features:

- **Maintaining Group Focus:** Private calls allow for quick and targeted communication between individuals without derailing the group conversation. Sensitive information or tactical maneuvers can be discreetly communicated without broadcasting them to the entire team.

- **Increased Efficiency:** Direct calling eliminates the need to announce call signs or wade through multiple conversations to reach a specific member. This expedites communication flow within the group.

Remember:

- **Compatibility:** Ensure all GMRS radios within your group are compatible with the chosen group call or private call features. Not all radios offer these functionalities, and some might have limitations on the number of group IDs or contacts that can be programmed.
- **Group Size Limitations:** The effectiveness of group calls can be impacted by group size. Extremely large groups might experience congestion or difficulty understanding conversations with multiple voices transmitting simultaneously. In such scenarios, consider dividing the group into smaller sub-groups for improved communication clarity.

By understanding and utilizing group call and private call features effectively, you can transform your GMRS radio into a powerful tool for coordinating and communicating seamlessly with your team during adventures, events, or any situation that demands efficient group interaction.

6.6 Unleashing the Potential: Data Transmission and Packet Radio for Tech-Savvy GMRS Users

While voice communication reigns supreme in the realm of GMRS radios, a hidden world of data transmission and packet radio techniques awaits those who crave to explore the potential beyond voice. This section probes the exciting possibilities for users with a technical background and a thirst for venturing beyond the realm of spoken words.

Embracing Data Transmission on GMRS Radios:

Imagine sending short text messages, sharing GPS coordinates with your team, or even transferring basic files between GMRS radios. While not all GMRS radios offer this functionality, some advanced models do. Here's what you need to know to unlock this capability:

- **Technical Knowledge is Key:** Successfully transmitting data on GMRS radios requires a solid grasp of technical concepts like baud rate (data transfer speed), error correction protocols (ensuring data integrity), and data modes (communication formats). Familiarize yourself with these terms and their impact on data transmission effectiveness. Resources like online tutorials, user manuals, and amateur radio forums can be invaluable assets in this journey.
- **Understanding the Limitations:** While data transmission opens doors, it's crucial to be aware of the inherent limitations of GMRS radios for this purpose.
    - **Range Restrictions:** Unlike dedicated data networks, GMRS radios have a shorter range for data transmission. Factors like terrain, distance, and obstacles can significantly impact the reach of your data packets.
    - **Slower Speeds:** Don't expect blazing-fast data transfer rates. GMRS data transmission typically operates at slower speeds compared to cellular networks or Wi-Fi. Be prepared for some patience when transferring larger files.
    - **Licensing Considerations:** In some regions, there might be additional licensing requirements for using GMRS radios for data transmission. Always consult local regulations to ensure you're operating within legal boundaries.

Gearing Up for Data Transmission:

Assuming your GMRS radio boasts data transmission capabilities and you've familiarized yourself with the technical aspects, here's how to get started:

1. **Consult Your Radio's Manual:** The user manual is your bible. It will provide detailed instructions on configuring data transmission settings on your specific radio model. This typically involves parameters like baud rate, error correction protocol, and data mode selection.
2. **Find a Compatible Partner:** Data transmission requires a compatible recipient radio equipped with data capabilities and configured with the same settings as yours. Coordinate with your partner(s) to ensure your radios are "speaking the same language" for successful data exchange.

3. **Software Might Be Required:** Some GMRS radios might require additional software installation on a connected computer to facilitate data transmission. The software acts as a bridge between your computer and the radio, enabling data transfer and configuration management.
4. **Test and Refine:** Once configured, don't be afraid to experiment! Start with short test transmissions at varying distances to assess the range limitations of your setup. Refine your settings based on your findings to optimize data transfer success.

Packet Radio: Diving Deeper (For the Truly Enthusiastic)

For those who crave an even deeper technical challenge, packet radio beckons. Packet radio utilizes digital packet switching techniques to transmit data over radio frequencies. If your GMRS radio supports data transmission, it might be configurable to function as a packet radio device, opening doors to exciting possibilities:

- **The Power of Packets:** Data is broken down into smaller, more manageable packets containing addressing information, error correction codes, and the actual data payload. This allows for efficient routing and retransmission if errors occur during transmission.
- **Packet Switching Magic:** Unlike traditional phone calls that establish a dedicated channel, packet switching routes each data packet independently through the network, taking the most efficient path to reach its destination. This allows for more efficient use of the shared GMRS frequency band.
- **The Allure of Packet Radio Networks:** A network of packet radio users or dedicated packet radio infrastructure can significantly extend the communication range and offer access to resources like electronic bulletin board systems (BBS) or even basic email services over radio waves!

Important Considerations for Packet Radio:

- **Advanced Technical Expertise is Mandatory:** Successfully navigating the world of packet radio requires a strong foundation in data communications, radio technology, and troubleshooting potential issues related to network configuration and packet protocols. Be prepared for a steeper learning curve compared to basic data transmission.

- **Limited Availability and Usage:** Packet radio usage on GMRS frequencies might be restricted in some regions due to regulations or the lack of established packet radio networks in your area. Research local regulations and explore online communities to assess the viability of packet radio in your location.

Conclusion: Data and Packet Radio - A Path for the Committed

Data transmission and packet radio techniques offer a glimpse into the fascinating potential of GMRS radios beyond basic voice communication. However, venturing into this territory requires dedication and a thirst for technical knowledge. Here are some parting thoughts to guide you on this path:

- **Start Slow and Build Confidence:** Don't be overwhelmed. Begin with mastering basic data transmission techniques like sending text messages between compatible radios. Once you're comfortable, gradually explore more complex functionalities like file transfer.
- **The Online Community is Your Friend:** The internet is a treasure trove of information for data-savvy GMRS users. Online forums, user groups, and websites dedicated to amateur radio and packet radio offer invaluable resources like troubleshooting guides, configuration tips, and discussions with experienced users.
- **Safety and Legality Come First:** Always prioritize safe operating practices within the legal boundaries set by your local regulations. Ensure you have the proper licenses if required for data transmission in your area. Respect the shared nature of the GMRS frequency band and avoid interfering with other users.
- **The Reward is in the Journey:** The process of learning and exploring data transmission and packet radio can be just as rewarding as the applications themselves. Embrace the challenge, enjoy the problem-solving process, and celebrate your successes along the way.

Note: For most GMRS users, the core strength lies in clear and reliable voice communication. Data transmission and packet radio are exciting avenues for those seeking to push the boundaries and explore the technical potential of their radios. If you're unsure, prioritize mastering the fundamentals of voice communication first. The world of data and packet radio will always be there waiting for you when you're ready to explore further.

Summary

This chapter ventured beyond the basics of GMRS radio communication, exploring a range of advanced features designed to enhance your experience and expand the functionality of your radio. We looked into:

- **Scanning Modes and Features:** Techniques for efficiently searching for active channels and conversations on the GMRS frequency band.
- **Dual Watch and Priority Scan:** Strategies for monitoring multiple channels simultaneously, ensuring you don't miss important messages.
- **CTCSS and DCS Codes:** Unveiling the world of subaudible tones that offer a layer of privacy and reduce channel clutter.
- **Weather Channels and Alerts:** Harnessing the power of NOAA Weather Radio for real-time weather information and critical weather alerts directly on your GMRS radio.
- **Multi-User and Group Communication:** Exploring group call and private call features to streamline communication and coordination within your team.

For technically-inclined users, we peeked into the fascinating realm of:

- **Data Transmission:** Transferring small digital data packets like text messages, GPS coordinates, or basic files between GMRS radios with data capabilities.
- **Packet Radio:** A more advanced concept utilizing digital packet switching techniques for communication over radio frequencies, potentially enabling connection to packet radio networks (availability and regulations permitting).

Always remember, effectively using these advanced features requires consulting your specific GMRS radio's user manual for detailed instructions. While data transmission and packet radio offer exciting possibilities, prioritize mastering the core functionalities of voice communication first. For most users, the beauty of GMRS radios lies in their ability to facilitate clear and reliable communication, and the advanced features explored in this chapter serve to enhance that core strength.

Review Questions

1. Describe the two main benefits of using CTCSS or DCS codes on your GMRS radio and explain the difference between how they function.
2. Imagine you're leading a group hike with several other GMRS radio users. Which features explored in this chapter would be most beneficial for your communication needs, and why?
3. Are you interested in exploring data transmission capabilities on your GMRS radio? Briefly explain the key considerations you would need to address before attempting data transmission.

CHAPTER 7

GMRS RADIO MAINTENANCE AND TROUBLESHOOTING

Your GMRS radio is a vital tool for communication and safety in the outdoors. Just like any dependable companion, it deserves proper care and attention to function at its best. This chapter equips you with the knowledge and resources to maintain your GMRS radio, troubleshoot common issues, and explore options for maximizing its performance.

Whether you're a seasoned outdoor enthusiast or a newcomer to the world of GMRS communication, this chapter is your one-stop guide to ensuring your radio remains a reliable partner in your adventures. We'll discuss:

- **Routine Maintenance Tips:** Simple yet effective practices to extend the lifespan and optimize the performance of your GMRS radio.
- **Troubleshooting Common Issues:** Learn how to diagnose and resolve some of the most frequently encountered problems with your radio.
- **Repair and Service Options:** Explore solutions for addressing more complex issues that might require professional repair services.
- **Upgrading and Enhancing Performance:** Consider potential upgrades to your radio or accessories for expanded functionality or improved communication capabilities (with licensing considerations in mind).
- **Spectrum Analyzers and Diagnostic Tools:** For the technically-inclined, we'll briefly explore advanced tools used for in-depth signal analysis and troubleshooting.

By following the practices and insights presented in this chapter, you can transform your GMRS radio from a simple communication device into a well-maintained and reliable tool, ready to serve you dependably whenever you venture into the great outdoors.

7.1 Routine Maintenance: Optimizing Performance and Extending the Lifespan of Your GMRS Radio

Effective maintenance is crucial for ensuring your GMRS radio operates reliably when you need it most. Here are practical tips to incorporate into your routine, maximizing performance and lifespan:

- **Regular Cleaning:** Dust, dirt, and debris can accumulate on the antenna and exterior of your GMRS radio, potentially impacting signal reception and overall functionality. Regularly wipe down your radio with a soft, damp cloth (avoiding excessive moisture) to remove dust and grime. Refrain from using harsh chemicals or abrasive cleaners that could damage the delicate surfaces.
- **Battery Care:** The battery is the lifeblood of your GMRS radio. To optimize battery life, follow these guidelines:
  - **Proper Storage:** Avoid storing your radio or spare batteries in extreme temperatures. Both hot and cold environments can degrade battery health and shorten lifespan. Opt for cool, dry storage conditions.
  - **Maintain Battery Health:** Even with infrequent use, perform periodic discharge and recharge cycles for your batteries. This helps maintain battery health and prevent memory effect, which can reduce overall capacity.
  - **Invest in Quality Replacements:** When replacing batteries, choose high-quality options from reputable brands. Avoid counterfeit batteries that could pose safety risks or damage your radio.
- **Safe Handling:** Accidents can happen, but treat your GMRS radio with care. Avoid dropping it, exposing it to excessive moisture, or subjecting it to rough handling. Consider using a sturdy carrying case for protection during transport and storage.
- **Review and Update Settings:** Regularly review your radio's settings to ensure they are configured correctly for your intended use. This includes verifying channel selections, privacy codes (CTCSS/DCS), and any other user-defined settings. Additionally, if available for your radio model, firmware updates can address bugs or introduce new features. Consult your user manual for instructions on accessing and updating settings or firmware.

Implementing these simple yet effective maintenance practices significantly improve the longevity and reliability of your GMRS radio, ensuring it remains a dependable communication tool for all your outdoor activities.

7.2 Troubleshooting Common GMRS Radio Issues

Even with proper maintenance, occasional glitches or unexpected behavior can occur with your GMRS radio. This section equips you with the knowledge to troubleshoot some of the most frequently encountered problems:

7.2.1 No Power or Low Battery:

- **Verify Battery Level:** The most common culprit for a powerless radio is a depleted battery. Replace the battery with a fully charged one or connect your radio to a power source if it supports external charging. Ensure the battery contacts are clean and free of corrosion.
- **Power Switch:** Double-check that the power switch is turned on completely. Sometimes, a loose connection or an incomplete rotation of the switch can mimic a dead battery.
- **External Power Source:** If your radio supports external power, try using a different power cable or power source to rule out issues with the charging mechanism.

7.2.2 Poor Reception or Range:

- **Antenna Check:** The antenna plays a critical role in good reception and transmission. Ensure the antenna is fully extended and upright. Inspect the antenna for any visible damage like cracks or loose connections.
- **Minimize Obstructions:** Walls, trees, and other large objects can significantly weaken the signal. Try moving to a location with a clearer line of sight to the intended recipient. If using a handheld radio, consider lowering the antenna slightly to avoid brushing against foliage.
- **Channel Confirmation:** Verify that you're on the correct channel for communication. If unsure, consult your user manual or local GMRS channel chart.

- **Range Limitations:** Remember that GMRS radios have a limited range. If you're experiencing poor reception beyond the expected range, consider using a repeater (if available in your area) or switching to a different communication method.

7.2.3 Audio Quality Problems:

- **Volume Adjustment:** Ensure the volume control is turned up to an audible level.
- **Speaker/Microphone Cleaning:** Dirt or debris can accumulate on the speaker and microphone grills, impacting audio quality. Gently clean the grills with a soft brush. If the issue persists, the speaker or microphone might be malfunctioning and require professional repair.
- **Signal Strength:** Weak signal strength can result in distorted or choppy audio. Refer to troubleshooting tips for poor reception (mentioned earlier) to improve signal strength.
- **CTCSS/DCS Compatibility:** Verify that both radios involved in the communication are programmed with the same CTCSS or DCS code (if applicable). Incompatible codes can cause garbled audio or complete lack of communication.
- **Interference:** Electrical interference from power lines or other electronic devices can sometimes introduce static or noise into the audio signal. Try moving away from potential sources of interference to see if the audio quality improves.

7.2.4 Programming Errors

- **Consult the Manual:** Programming errors can lead to unexpected behavior. Refer to your radio's user manual for detailed instructions on programming channels, privacy codes, and other settings. Many user manuals are also available online in PDF format for easy reference.
- **Double-Check Entries:** Carefully review your programming entries for any typos or incorrect selections. A single misplaced number or letter can throw off your radio's functionality.
- **Reset to Factory Settings:** If all else fails, consider resetting your radio to factory settings. This will erase all custom programming and restore the radio to its default configuration. Consult your user manual for the specific steps involved in performing a factory reset. Remember, a factory reset will erase all your custom

settings, so proceed with caution and ensure you have backups of important information like channel frequencies or privacy codes before resetting.

By systematically working through these troubleshooting steps, you can often resolve common issues with your GMRS radio and restore it to full functionality. If the problem persists after trying these solutions, refer to section 7.3 for information on repair and service options.

7.3 Repair and Service Options: Addressing Complex Issues with Your GMRS Radio

Sometimes, even the best troubleshooting efforts might not resolve an issue with your GMRS radio. In such cases, professional repair services can help get your radio back in working order. Here's a breakdown of your options:

- **Warranty Coverage:** If your radio is still under warranty, check the warranty terms and conditions. If the issue falls under warranty coverage, contact the manufacturer or an authorized dealer for repair or replacement options.

- **Authorized Repair Centers:** Many electronics manufacturers have a network of authorized repair centers equipped with the expertise and tools to diagnose and service their products. Locate an authorized repair center in your area for professional service.
- **Independent Repair Shops:** Depending on the complexity of the issue and their experience with specific radio models, independent electronics repair shops might also be able to service your GMRS radio.

Important Considerations:

- **Cost-Effectiveness:** Before opting for repairs, weigh the repair cost against the value of your radio. For older radios, the cost of repair might exceed the value of replacing it with a newer model.
- **Data Backup:** If your radio model allows storing data like programmed channels, privacy codes, or contact information, ensure you have backups of this data before sending your radio for repair. The repair process might erase these settings.

Choosing the Right Repair Option:

The best repair option depends on several factors:

- **Warranty Status:** If under warranty, authorized repair through the manufacturer is typically the most recommended course of action.
- **Complexity of the Issue:** For complex issues requiring specialized tools or expertise, authorized repair centers might be the best choice.
- **Cost Considerations:** Independent repair shops can sometimes offer more affordable repair options, but ensure they have the experience and qualifications to service your specific radio model effectively.

A careful evaluation of these factors will empower you to select the most suitable repair service for your GMRS radio. This guarantees it receives the necessary attention to return to full functionality.

7.4 Upgrading and Enhancing Performance: Expanding the Capabilities of Your GMRS Radio

As your needs or interests in outdoor communication evolve, you might consider upgrading your GMRS radio for enhanced performance or additional features. Here's a breakdown of some factors to consider when contemplating an upgrade:

- **New Features and Technologies:** Newer GMRS radio models might offer advancements like improved audio quality, extended battery life, larger channel capacities, or compatibility with digital communication modes (if future-proofed for such advancements). Researching the latest features on the market can help you determine if an upgrade would significantly benefit your communication needs.
- **Increased Range and Power (with Licensing Considerations):** For users who obtain a GMRS repeater license, higher-powered radios can be used in conjunction with repeaters to significantly extend communication range. It's crucial to ensure you comply with all local regulations regarding GMRS radio power output and licensing requirements before considering such an upgrade.
- **Accessories for Improved Functionality:** Investing in accessories like external antennas, headsets, or speaker microphones can enhance the usability and performance of your existing GMRS radio. Consider your specific communication needs and preferences when exploring these options.

Important Considerations:

- **Cost vs. Benefit:** Carefully weigh the cost of upgrading your radio against the potential benefits it offers. Consider how the new features or functionalities align with your current and future communication requirements.
- **Compatibility:** Ensure any accessories you purchase, such as external antennas or microphones, are compatible with your specific GMRS radio model. Consult your user manual or the manufacturer's website for compatibility information.

Upgrading your GMRS radio can be a strategic decision, allowing you to leverage the latest advancements in technology or expand the communication capabilities of your existing radio. A thoughtful assessment of your needs alongside the factors discussed above will equip you to decide whether upgrading your GMRS radio is the best course of action.

7.5 Spectrum Analyzers and Diagnostic Tools: Advanced Troubleshooting for the Technically Inclined

For users who are comfortable with deeper dives into troubleshooting or have a strong technical background, spectrum analyzers and diagnostic tools can be valuable assets in their GMRS radio toolkit. These tools provide advanced capabilities beyond basic troubleshooting methods:

- **Visualizing Signal Strength:** Spectrum analyzers offer a graphical representation of the radio frequency spectrum. This allows you to identify potential sources of interference that might be impacting your radio's performance. Visualizing the signal strength can also aid in diagnosing issues related to signal quality or transmission power.
- **Advanced Diagnostics:** Certain diagnostic tools can provide detailed information about the internal functions of your GMRS radio. This can be helpful in pinpointing complex problems that might not be readily apparent through basic troubleshooting steps. For example, these tools might offer insights into component functionality, signal processing details, or internal error codes.

Important Considerations:

- **Technical Expertise:** Using spectrum analyzers and diagnostic tools effectively requires a strong understanding of radio technology and signal analysis principles. An in-depth knowledge of GMRS radio operation and communication protocols is also beneficial for interpreting the data obtained from these tools.
- **Cost and Investment:** Spectrum analyzers and diagnostic tools can be expensive pieces of equipment. For casual GMRS radio users, the cost of these tools might not be justified.

If you're a technically-inclined user who wants to go deeper into troubleshooting or optimize your GMRS radio's performance to the fullest potential, then investing in spectrum analyzers and diagnostic tools might be a worthwhile consideration. However, for most users, a basic understanding of troubleshooting techniques covered earlier in this chapter will likely be sufficient to address most common GMRS radio issues.

Summary

This chapter transformed your GMRS radio from a simple communication device into a reliable companion for your outdoor adventures. We covered essential practices for maximizing its performance and lifespan:

- **Routine Maintenance:** Regularly cleaning your radio, caring for the battery, handling it with respect, and keeping settings updated ensures smooth operation.
- **Troubleshooting Common Issues:** A step-by-step approach equips you to address common problems like no power, poor reception, audio quality issues, and programming errors.
- **Repair and Service Options:** If troubleshooting fails, explore options like warranty repairs, authorized service centers, or independent repair shops, considering factors like warranty status, complexity of the issue, and cost-effectiveness.
- **Upgrading for Enhanced Performance:** Evaluate the need for an upgrade based on factors like new features, increased range with proper licensing, or accessory compatibility to improve functionality.
- **Advanced Tools for Tech Users (Optional):** For those comfortable with technical aspects, spectrum analyzers and diagnostic tools offer advanced troubleshooting capabilities for visualizing signal strength and performing in-depth diagnostics.

Adhering to the guidelines provided in this chapter ensures your GMRS radio remains a reliable communication device, prepared for all your outdoor adventures. It's important to note that basic maintenance and troubleshooting methods are typically sufficient to resolve most common problems. However, for those with a more technical background, exploring advanced tools can offer deeper insights into enhancing performance.

Review Questions

1. Describe two routine maintenance practices you should perform to extend the life and optimize the performance of your GMRS radio battery.
2. You're on a hiking trip and experience crackling audio on your GMRS radio. List two possible causes of this issue and explain the troubleshooting steps you would take to identify the culprit.
3. When considering an upgrade to a newer GMRS radio model, what are three key factors you should weigh in your decision-making process (according to the information presented in this chapter)?

CHAPTER 8

GMRS RADIO SETTINGS AND CONFIGURATION - MASTERING YOUR COMMUNICATION HUB

Your GMRS radio is more than just a communication device; it's a customizable tool tailored to your specific needs. This chapter explores the exciting world of GMRS radio settings and configuration, empowering you to unlock the full potential of your radio. Whether you're a seasoned outdoor enthusiast or a newcomer to GMRS communication, this chapter equips you with the knowledge and skills to:

- **Fine-tune basic settings** like squelch and volume levels to optimize audio clarity and reception for your environment.
- **Configure channel scanning** to efficiently search for active channels and ensure you stay connected with your group.
- **Customize alert tones and alarms** to personalize your radio's notifications and enhance situational awareness.
- **Explore firmware updates and software customization options** (if available for your radio model) to keep your radio functioning at its best and potentially unlock new features.
- **Understand radio system integration and interfacing** (applicable for advanced users) to explore possibilities of connecting your GMRS radio with other communication systems.

Equipped with the knowledge to navigate these settings and configurations, you'll be empowered to elevate your GMRS radio from a basic communication tool to a personalized and versatile companion for all your outdoor adventures. Let's embark on this journey of optimizing your GMRS radio and maximizing its capabilities!

8.1 Basic Settings Overview: Optimizing Your GMRS Radio for Everyday Use

Every GMRS radio comes equipped with a core set of user-configurable settings that significantly impact your communication experience. Familiarizing yourself with these basic settings allows you to fine-tune your radio's operation for optimal performance in various environments. Here's a breakdown of some of the most common settings you'll encounter:

- **Channel Selection:** GMRS radios operate on a designated set of frequencies divided into channels. This setting allows you to choose the specific channel you want to use for communication. Consulting your local GMRS channel chart or user manual will help you identify the appropriate channel for your area and intended use.

- **Squelch Control:** Squelch is a noise reduction feature that mutes the speaker output when no signal is present. This helps eliminate background noise and static when no one is transmitting on your selected channel. Adjusting the squelch level allows you to strike a balance between eliminating excessive noise and ensuring you don't miss weak signals.
- **Volume Control:** This setting adjusts the overall speaker volume of your GMRS radio. It's crucial to set a comfortable listening level that allows you to hear incoming transmissions clearly, especially in noisy environments.
- **CTCSS/DCS Codes (Optional):** These are subaudible tones that can be used for added privacy and to reduce channel congestion. By programming the same CTCSS or DCS code on both your radio and the radios of your communication group, you can ensure you only hear transmissions from within your group and minimize interference from other users on the same channel. Not all GMRS radios offer CTCSS/DCS functionality, so consult your user manual for details.
- **Roger Beep/Auto Roger (Optional):** A Roger beep is a brief tone transmitted at the end of your transmission to indicate you've finished speaking. Auto Roger automatically transmits this tone upon releasing the PTT (Push-to-Talk) button. These features are optional and depend on user preference and communication etiquette within your group.

Understanding and effectively utilizing these basic settings empowers you to optimize your GMRS radio for clear communication, minimize distractions from background noise, and potentially enjoy a layer of privacy with CTCSS/DCS codes (if your radio supports them). The following sections will plunge into specific functionalities like squelch adjustment, channel scanning configuration, and customization options to further enhance your GMRS radio experience.

8.2 Adjusting Squelch and Volume Levels: Striking the Right Balance

Imagine this: you're on a thrilling hike with your friends, relying on your GMRS radios to stay connected. Suddenly, a constant hiss fills the speaker, making it difficult to hear incoming messages. This unpleasant scenario highlights the importance of fine-tuning your squelch level. Let's explore how to adjust squelch and volume for optimal audio clarity:

- **Squelch Control:** As mentioned earlier, squelch mutes the speaker output when no signal is present. A properly adjusted squelch level eliminates excessive background noise and static but ensures you don't miss weak signals. Here's how to find the sweet spot:
    1. **Locate the Squelch Knob:** Consult your user manual to identify the squelch control knob or button on your radio.
    2. **Start with Minimum Squelch:** Turn the squelch knob to its minimum setting (usually counter-clockwise). In this state, you'll likely hear significant background noise.
    3. **Gradually Increase Squelch:** Slowly turn the squelch knob clockwise until the background noise subsides. The ideal setting is the point where you hear minimal to no noise when no signal is present, but weak incoming transmissions are still audible.
- **Volume Control:** Setting an appropriate volume level is crucial for clear communication, especially in noisy environments. Here's a simple approach:
    1. **Adjust in a Quiet Environment:** If possible, find a quiet area to adjust the volume initially.
    2. **Set a Comfortable Level:** Turn the volume control up gradually until you can hear transmissions clearly and comfortably. Avoid excessively high volumes that can lead to discomfort or ear fatigue.
    3. **Fine-Tune Based on Environment:** Be prepared to adjust the volume throughout your adventure as background noise levels change.

With some practice adjusting the controls in various settings, you'll soon become adept at achieving optimal audio clarity for your GMRS radio communications. Now, let's explore channel scanning, a valuable feature for efficiently searching for active channels on your GMRS radio.

8.3 Channel Scanning Configuration: Finding Your Frequency with Ease

Imagine you're out exploring a new trail and need to connect with your group. Manually scanning through all the available GMRS channels can be time-consuming and inefficient. This is where channel scanning comes in – a convenient feature that allows your radio to automatically search for active channels within the GMRS frequency range. Here's a breakdown of channel scanning and how to configure it:

- **Understanding Channel Scanning:** When activated, channel scanning instructs your radio to rapidly cycle through all GMRS channels, briefly stopping on each one to check for activity (i.e., ongoing transmissions). If the radio detects a signal on a channel, it will typically emit a beep or display an indicator on the screen, allowing you to identify potentially active channels for communication.
- **Configuring Channel Scanning:** The specific steps for configuring channel scanning will vary depending on your GMRS radio model. However, most radios follow a similar approach:
  1. **Consult User Manual:** Refer to your user manual for detailed instructions on accessing the channel scanning menu on your specific radio model.
  2. **Locate Scan Options:** Look for settings related to channel scanning, which might be labeled as "Scan," "Search," or similar terms.
  3. **Priority Channel Selection (Optional):** Some radios allow you to designate a priority channel that the scan will pause on for a longer duration if activity is detected. This can be helpful if you have a pre-arranged communication channel with your group.
  4. **Initiate Scan:** Once you've configured the desired settings, locate the button or function to initiate the channel scan.
- **Utilizing Scan Results:** After the scan is complete, your radio will typically display a list of channels with detected activity. You can then manually select the most relevant channel for your communication needs.

To significantly reduce the time spent searching for active channels, especially in areas with many GMRS radio users, it is essential to effectively utilize channel scanning. This feature is particularly beneficial when coordinating communication with your group in unfamiliar locations or during events with high traffic on GMRS channels. By leveraging channel scanning, you can efficiently navigate through the frequency spectrum, identifying active channels with minimal effort. This not only streamlines your communication process but also enhances your ability to connect with your group in various settings, ensuring seamless and effective communication.

The next section will look into customizing alert tones and alarms on your GMRS radio, allowing you to personalize your communication experience and enhance situational awareness.

8.4 Customizing Alert Tones and Alarms: Personalizing Your Radio Experience

Your GMRS radio isn't just for voice communication; it can also provide vital auditory cues through alert tones and alarms. These features not only enhance your situational awareness but also allow for a degree of personalization to your radio experience. Let's explore the world of customizing alert tones and alarms on your GMRS radio:

- **Alert Tone Options:** Many GMRS radios offer a variety of pre-programmed alert tones that can be used for different purposes. These tones can notify you of incoming calls, low battery levels, or the completion of a channel scan. Here are some common alert tones and their potential uses:
    - o **Call Alert:** A distinct tone to signal an incoming transmission on your selected channel.
    - o **Roger Beep:** A brief tone you can configure to transmit automatically upon releasing the PTT button, indicating the end of your transmission (optional functionality, etiquette may vary).
    - o **Low Battery Alert:** A warning tone to notify you when your battery level dips below a certain threshold.
    - o **Channel Scan Complete:** An audible cue indicating the radio has finished scanning all available channels.
- **Customizing Alert Tones:** Some GMRS radios allow you to select your preferred alert tones from a pre-programmed list. Consult your user manual to explore the available options and customize your notification sounds for a more personalized experience.
- **Alarm Functionality (Optional):** Certain GMRS radios offer programmable alarms that can be used for various purposes, such as setting reminders or acting as a safety beacon. The specific functionalities and configuration steps will vary depending on the model, so refer to your user manual for detailed instructions.

To significantly enhance your GMRS radio experience, it is essential to effectively utilize and customize alert tones and alarms. Distinctive call alerts are designed to ensure that you do not miss any important messages, thereby ensuring that your communication remains uninterrupted. Additionally, low battery warnings are crucial as they prevent unexpected radio shutdowns, ensuring that your communication device remains operational at all times. Furthermore, the optional alarm functionalities can be tailored to meet your specific needs, adding an extra layer of safety or convenience to your radio experience. This customization allows you to personalize your GMRS radio to better suit your requirements, enhancing both your safety and convenience.

The next section explores firmware updates and software customization, which can unlock new features or improve the overall functionality of your GMRS radio (depending on the model and manufacturer).

8.5 Firmware Updates and Software Customization: Keeping Your Radio Up-to-Date (if applicable)

Technology is constantly evolving, and the world of GMRS radios is no exception. Some radio models offer firmware updates and software customization options that can enhance functionality, address bugs, or even introduce new features. Let's explore these possibilities, keeping in mind that not all GMRS radios will have these capabilities.

- **Understanding Firmware Updates:** Firmware is the underlying software that controls the operation of your GMRS radio. Manufacturers occasionally release firmware updates that can address bugs, improve performance, or introduce new functionalities.
- **Benefits of Firmware Updates:** Installing a firmware update can potentially:
    - Fix bugs or glitches you might have encountered while using your radio.
    - Enhance the overall performance and stability of your radio.
    - In rare cases, introduce entirely new features or functionalities not available in the original firmware.
- **Software Customization (Optional):** A limited number of GMRS radios offer software customization options that allow you to personalize settings beyond the basic configurations discussed earlier. These customizations might involve:

- Programming additional channels or channel names for easier identification.
- Adjusting advanced operational parameters (for experienced users with a strong understanding of GMRS radio technology).
- **Checking for Updates:** The process for checking for and installing firmware updates will vary depending on your specific GMRS radio model. Here's a general approach:
  - **Consult the User Manual:** Refer to your user manual for instructions on how to check for available firmware updates for your radio model. Look for sections related to "firmware updates," "software updates," or similar terms.
  - **Manufacturer's Website:** Many manufacturers provide information about firmware updates on their websites. Visit the website of the company that manufactured your GMRS radio and search for your specific model to see if updates are available.

Important Considerations:

- **Not All Radios Have This Capability:** As mentioned earlier, not all GMRS radios offer firmware update or software customization options. Consult your user manual to confirm if your radio supports these features.
- **Proceed with Caution:** While firmware updates can be beneficial, it's crucial to follow the manufacturer's instructions carefully during the update process. Installing an update incorrectly could potentially damage your radio.
- **Only Use Approved Updates:** Ensure you're downloading and installing firmware updates from a trusted source, preferably the manufacturer's website. Avoid installing updates from unverified third-party sources.

To maintain your GMRS radio's up-to-date status and to reap the benefits of the latest performance enhancements or bug fixes, it is crucial to stay informed about and apply available firmware updates, provided they are applicable to your radio model. This proactive approach ensures that your radio remains at the forefront of technological advancements. The concluding segment of this chapter will explore the intricacies of radio system integration and interfacing. This section is specifically designed for more advanced users who are interested in the process of connecting their GMRS radio with other communication systems, thereby expanding their communication capabilities.

8.6 Radio System Integration and Interfacing: Advanced Communication Possibilities (for Experienced Users)

This chapter has equipped you with the knowledge to configure your GMRS radio for optimal performance and personalize your communication experience. For the truly tech-savvy users out there, this final section explores the advanced topic of radio system integration and interfacing. Here, we'll explore the possibilities of connecting your GMRS radio with other communication systems, expanding its capabilities beyond basic GMRS channel communication.

Understanding Radio System Integration:

Radio system integration involves connecting your GMRS radio with other communication systems to create a more complex network. This can offer advantages in specific scenarios, such as:

- **Large-Scale Event Management:** Organizers of events with a wide coverage area might integrate GMRS radios with repeaters or other communication systems to extend range and coordinate activities across a larger geographical area.
- **Interoperability with Different Radio Systems (Advanced):** In some situations, it might be desirable to connect your GMRS radio with other radio systems, such as amateur (ham) radio or commercial two-way radio systems (with proper licensing and permissions). However, achieving interoperability can be complex and requires a deep understanding of different radio technologies and protocols.

Interfacing Methods (Advanced):

There are various methods for interfacing your GMRS radio with other systems, depending on the desired functionality and the specific equipment involved. Here are a few general possibilities:

- **Cable Connections:** Certain radios offer dedicated ports for connecting cables that interface with other communication systems. These cables might require specific configurations or protocols to function properly.

- **Digital Mode Integration (Advanced):** Some advanced GMRS radios support digital communication modes that can be interfaced with other digital radio systems, potentially offering improved audio quality and additional features.

Important Considerations:

- **Technical Expertise Required:** Radio system integration and interfacing are advanced topics that require a strong understanding of radio technology, communication protocols, and potentially the specific equipment involved. If you're new to GMRS radios, it's recommended to focus on mastering the basic functionalities covered in this chapter before venturing into this complex area.
- **Licensing and Regulations:** Depending on the type of interfacing you attempt and the radio systems involved, additional licensing or authorization might be required by law. Ensure you comply with all relevant regulations before attempting any advanced integration projects.

Conclusion:

For most GMRS radio users, the functionalities covered in the preceding sections of this chapter will likely be sufficient for effective communication. However, for the technically inclined users who crave more complex communication solutions, radio system integration and interfacing offer exciting possibilities. Remember to approach this advanced topic with caution, ensure you possess the necessary technical expertise, and comply with all relevant regulations.

By familiarizing yourself with the settings, configurations, and advanced functionalities explored in this chapter, you've transformed your GMRS radio from a simple communication tool into a versatile companion for all your outdoor adventures. Now, you possess the knowledge to optimize its performance, personalize your experience, and potentially explore the exciting world of radio system integration (for experienced users). Happy trails and clear communication!

Summary

This chapter empowered you to transform your GMRS radio from a basic device into a personalized and versatile communication tool for your outdoor adventures. We explored a range of settings and configurations to optimize your radio's performance:

- **Fine-tuning Basics:** You learned to adjust squelch and volume levels for optimal audio clarity in various environments.
- **Channel Efficiency:** Channel scanning was introduced as a helpful feature for efficiently searching for active channels among those available on your GMRS radio.
- **Customization Options:** Explore alert tone and alarm functionalities to personalize your radio experience and enhance situational awareness. Some models even allow customization of pre-programmed tones.
- **Staying Up-to-Date (if applicable):** For radio models that offer firmware updates and software customization, we discussed the potential benefits of installing updates (improved performance, bug fixes, or new features) and offered guidance on how to check for their availability. Remember, not all radios have this capability, so consult your user manual for details.
- **Advanced Interfacing (for experienced users only):** The chapter concluded with a glimpse into the world of radio system integration and interfacing, a complex topic for technically-savvy users who wish to explore connecting their GMRS radio with other communication systems for extended range or interoperability with different radio types (proper licensing required).

Having gained a comprehensive grasp of these settings and configurations, you've successfully unlocked the full capabilities of your GMRS radio, thereby enabling clear, efficient communication, and even introducing a personalized element. It's crucial to remember that the functionalities outlined in this chapter are meticulously crafted to meet the needs of the vast majority of GMRS radio users. The segment focusing on advanced interfacing is specifically aimed at individuals with a solid technical background and a keen curiosity to dig into sophisticated communication solutions.

Review Questions

1. Describe two basic settings you can adjust on your GMRS radio to optimize audio clarity and explain how to adjust them for best results.
2. Imagine you're on a camping trip with a large group spread across several campsites within a few miles of each other. How could the channel scanning function on your GMRS radio help you locate the channel your group is using to communicate?
3. This chapter briefly mentioned firmware updates for GMRS radios. Explain two important considerations to keep in mind before installing a firmware update on your radio.

CHAPTER 9

GMRS RADIO BEST PRACTICES AND SAFETY - ENSURING RESPONSIBLE AND EFFECTIVE COMMUNICATION

Equipping yourself with the knowledge to operate your GMRS radio effectively is only half the journey. Responsible and safe communication practices are paramount for maximizing the utility of your GMRS radio while ensuring a positive experience for yourself and others on the shared GMRS channels. This chapter plunges into essential best practices and safety considerations that will elevate you from a casual user to a responsible and informed GMRS radio operator. Here, we'll explore key areas to focus on:

- **Safety Guidelines for Users:** We'll establish crucial safety protocols to ensure your GMRS radio usage doesn't put yourself or others at risk.
- **Interference Avoidance Techniques:** Learn how to identify and minimize interference that can disrupt your communication or that you might cause to others.
- **Environmental Considerations:** Understanding how environmental factors can impact your GMRS radio's performance allows you to adapt your communication strategies for various conditions.
- **Radio Security and Privacy Measures:** Explore steps you can take to protect the privacy of your communications and minimize the risk of unauthorized access to your GMRS radio.
- **Compliance with Regulatory Standards and Guidelines:** As a licensed GMRS radio user, it's essential to understand and adhere to the regulations governing GMRS communication to avoid potential legal repercussions.

Adopting these best practices and safety considerations is a crucial step towards contributing to a more positive and responsible communication environment for all GMRS radio users. This collective effort not only enhances the overall experience for each user but also fosters a safer and more harmonious communication space. Let's embark on this journey of becoming a safe and responsible GMRS radio operator together, paving the way for a more unified and effective communication community.

9.1 Safety Guidelines for GMRS Users

Your GMRS radio is a powerful tool for staying connected and coordinating activities during your outdoor adventures. However, the thrill of exploration shouldn't overshadow the importance of safety when relying on this technology. This section dives deep into essential safety guidelines that will transform you from a casual user into a responsible and informed GMRS radio operator:

- **Prioritizing Emergency Situations:** Always remember that GMRS is not a substitute for contacting emergency services. If you encounter a life-threatening situation, your first and most crucial action should be to dial 911 (or the appropriate emergency number in your region) immediately. However, GMRS can play a vital role in coordinating rescue efforts or requesting assistance in non-life-threatening emergencies, particularly in remote areas where cellular service might be unreliable. Think of it as a valuable backup communication tool, but never a primary one for critical situations.
- **Understanding Your Limitations:** Before embarking on any adventure, take a moment to assess your surroundings and acknowledge the limitations of your GMRS radio. Don't solely rely on GMRS communication for critical situations,

especially in areas with limited range due to geographical obstacles or potential signal obstructions. Always prioritize your safety by carrying additional safety equipment like a personal locator beacon (PLB) that can transmit a distress signal directly to emergency responders, even without a cellular connection. A PLB can be a lifesaver in situations where GMRS communication might be compromised.

- **Proactive Battery Management:** A dead battery in your GMRS radio can render it useless during an emergency. Develop the habit of checking battery levels regularly before and during your adventures. Don't be caught unprepared – carry spare batteries, especially on extended trips. Consider investing in a rechargeable battery and a reliable charging solution for frequent use. A fully charged radio is a silent guardian, ensuring you have a communication lifeline when you need it most.

- **Clarity and Concision During Critical Communication:** When faced with a critical situation, clear and concise communication is paramount. Speak slowly and articulate each word clearly. Identify yourself and your location precisely, allowing responders to pinpoint your situation. Avoid technical jargon or overly complex language that others might not understand. Focus on relaying vital information efficiently, keeping transmissions brief to prevent blocking the channel for others who might also need assistance. Remember, every second counts in an emergency, so make your communication clear, concise, and informative.

- **Respecting the Shared Communication Environment:** The GMRS channels are a valuable shared resource for everyone who relies on this technology for communication outdoors. Be mindful of the length of your transmissions and avoid lengthy conversations that could prevent others from using the channel for critical communication. Maintain a courteous tone and avoid offensive language. Think of it as a two-way street – use the channel responsibly to ensure everyone has the opportunity to communicate effectively.

Adhering to these safety guidelines ensures that your GMRS radio usage contributes to a safe and responsible communication environment for everyone on the shared channels. This collective effort not only enhances the overall experience for each user but also fosters a safer and more harmonious communication space. The next section will look into techniques for minimizing interference, a common challenge in GMRS communication. This exploration is crucial for understanding and addressing the various factors that can lead to interference, thereby improving the reliability and effectiveness of GMRS communication.

9.2 Interference Avoidance Techniques: Ensuring Clear Communication on the GMRS Channels

The beauty of GMRS radios lies in their ability to facilitate communication in remote locations. However, this shared environment can sometimes become congested, leading to a frustrating phenomenon called interference. Interference manifests as unwanted noise, crackling, or garbled audio that disrupts clear communication on your GMRS radio. This section equips you with valuable techniques to minimize interference and ensure your messages are received loud and clear:

- **Understanding Interference Sources:** The first step towards mitigating interference is understanding its potential sources. Here are some common culprits:
  - **Other GMRS Users:** Sharing the limited number of available channels can lead to overlapping transmissions, especially in areas with high GMRS radio activity. This can result in crosstalk or audio distortion.
  - **Electronic Devices:** Certain electronic devices, such as poorly shielded power lines, faulty electrical equipment, or even nearby smartphones, can emit electromagnetic waves that interfere with GMRS radio signals.
  - **Geographical Obstructions:** Mountains, dense forests, or buildings can act as barriers, weakening or deflecting GMRS radio signals, leading to reduced range and potential for interference, especially at the fringes of your radio's coverage area.
  - **Atmospheric Conditions:** In rare instances, even atmospheric conditions like solar flares or heavy thunderstorms can temporarily disrupt radio communication by introducing static or noise.
- **Minimizing Interference from Other Users:**
  - **Channel Etiquette:** Observe proper channel etiquette by keeping transmissions brief and to the point. Allow others a chance to communicate and avoid lengthy conversations that could clog the channel.

- o **Channel Selection:** If you experience interference on your current channel, try switching to a less congested one. Consult your local GMRS channel chart or user group to identify alternative channels with potentially lower traffic.
- o **Utilize CTCSS/DCS Codes (if available):** If your GMRS radio supports CTCSS or DCS codes (subaudible tones for added privacy and reduced congestion), consider programming the same code on your radio and the radios of your communication group. This can help minimize interference from other users on the same channel who are not using the same code.
- **Mitigating Interference from Electronic Devices:**
  - o **Identify the Culprit:** If you suspect a specific electronic device is causing interference, try turning it off or moving it further away from your GMRS radio to see if the issue resolves.
  - o **Invest in Shielded Equipment:** Consider using shielded cables for your GMRS radio antenna or microphone to minimize interference from nearby electronics.
- **Optimizing for Geographical Considerations:**
  - o **Strategic Positioning:** When possible, try to position yourself in a location with minimal obstructions between you and the person you're communicating with. Higher ground or areas with clear lines of sight can improve signal strength and reduce interference caused by geographical barriers.
  - o **Utilize Repeaters (if available):** In areas with significant geographical obstacles or limited range, consider using GMRS repeaters if available in your region. These are strategically placed stations that amplify and retransmit GMRS signals, extending your communication range and potentially bypassing local obstructions.
- **Adapting to Atmospheric Conditions:**
  - o **Limited Control:** Unfortunately, there's little you can do to control atmospheric interference caused by solar flares or severe weather events. However, being aware of these potential disruptions can help you manage expectations and consider alternative communication methods (e.g., pre-

arranged visual signals) if radio communication becomes unreliable during such events.

Understanding the sources of interference and implementing these techniques is a key strategy to significantly improve the clarity and reliability of your GMRS radio communication. This comprehensive approach not only enhances the quality of your communication but also contributes to a more efficient use of the spectrum. The next section will look into environmental considerations that can impact your GMRS radio's performance and how to adapt your communication strategies for various conditions. This exploration is crucial for understanding and addressing the various factors that can lead to interference, thereby improving the reliability and effectiveness of GMRS communication.

9.3 Environmental Considerations: Optimizing Your GMRS Radio for Different Conditions

Your GMRS radio is a dependable companion for outdoor adventures, but its performance can be influenced by various environmental factors. Understanding these factors and adapting your communication strategies accordingly allows you to maximize the effectiveness of your GMRS radio in diverse conditions. Here's a breakdown of some key environmental considerations:

- **Temperature Extremes:** Both extreme heat and cold can impact the performance of your GMRS radio battery. Hot temperatures can accelerate battery drain, while cold temperatures can reduce battery life and potentially hinder the functionality of the LCD screen or internal components.
  - o **Hot Weather Strategies:** During hot weather expeditions, consider carrying spare batteries or keeping your radio in a cool, shaded location when not in use. Opt for rechargeable batteries that tend to perform better in high temperatures compared to disposable alkaline batteries.
  - o **Cold Weather Strategies:** In cold weather environments, conserve battery life by keeping transmissions brief and minimizing the time your radio is powered on. Carry spare batteries and store them close to your body to

maintain warmth and optimal performance. Some manufacturers offer extended-life batteries specifically designed for cold weather use.

- **Precipitation:** Exposure to rain, snow, or dust can potentially damage your GMRS radio if not properly protected. While some radios boast weather-resistant features, it's always wise to exercise caution.
  - ○ **Weatherproof Considerations:** If your radio isn't explicitly advertised as weatherproof, invest in a carrying case or pouch that offers protection from rain, dust, and debris. Avoid submerging your radio in water, even if it claims some degree of water resistance.

- **Altitude:** As you gain altitude, the air thins, potentially affecting the range of your GMRS radio. While the impact might be negligible for small elevation changes, it's a factor to consider for high-altitude adventures.
  - ○ **Understanding Range Limitations:** Be familiar with the typical range of your GMRS radio under normal conditions. At higher altitudes, expect a potential decrease in range due to the thinner air. Plan your communication strategies accordingly, considering factors like terrain and potential obstacles that might further limit your effective communication range.
  - ○ **Repeaters as Allies (if available):** In high-altitude environments, utilizing GMRS repeaters (if available in your region) can significantly extend your communication range by amplifying and retransmitting your signal. Research the availability and locations of repeaters in your intended high-altitude adventure area.

- **Terrain:** Mountains, hills, and dense forests can act as barriers, weakening or deflecting GMRS radio signals. Understanding the terrain you'll be navigating can help you anticipate potential communication challenges.
  - ○ **Strategic Communication:** When possible, try to communicate from locations with clear lines of sight to the person you're trying to reach. Higher ground or areas with minimal obstructions between you can improve signal strength and overall communication effectiveness.

Acknowledging these environmental considerations and adapting your communication strategies accordingly ensures your GMRS radio remains a reliable tool for staying connected and ensuring safety during your outdoor adventures. This proactive approach not only enhances the effectiveness of your GMRS radio but also contributes to a safer and more efficient communication environment for all users. The next section will explore steps you can take to enhance the security and privacy of your GMRS radio communications. This exploration is crucial for understanding and addressing the various factors that can compromise the security and privacy of your communications, thereby ensuring that your GMRS radio remains a secure and private communication tool.

9.4 Radio Security and Privacy Measures: Protecting Your Communications

While GMRS radio communication offers a valuable tool for outdoor enthusiasts, it's important to acknowledge that these channels are not inherently secure. Anyone within range with a compatible GMRS radio can potentially listen to your transmissions. This section explores steps you can take to enhance the security and privacy of your GMRS radio communications:

- **Understanding the Limitations of GMRS Security:** GMRS channels operate on publicly accessible frequencies. There's no built-in encryption mechanism to scramble your voice communications, making them potentially vulnerable to interception by anyone within range with a GMRS radio tuned to the same channel.
- **Privacy Codes (CTCSS/DCS) -** (Optional Functionality): Some GMRS radios offer features like CTCSS (Continuous Tone-Coded Squelch System) or DCS (Digital Coded Squelch). These functionalities utilize subaudible tones that are inaudible to the human ear. By programming the same CTCSS/DCS code on your radio and the radios of your communication group, you can potentially minimize interference from other users on the same channel who are not using the same code. However, it's important to remember that CTCSS/DCS codes do not encrypt your communication; they simply act as a filter to reduce unwanted noise from other transmissions on the same channel. Someone with a GMRS radio deliberately set to scan for active channels could still potentially intercept your communication even if you're using CTCSS/DCS codes.
- **Strategic Communication Practices:**
  - **Minimize Sensitive Information:** Avoid transmitting highly sensitive information over your GMRS radio, especially if you suspect someone

might be listening. For truly private communication, consider alternative methods like encrypted messaging on cellular devices (if available) or pre-arranged visual signals.

- o **Use Code Words (Sparingly):** While code words can add a layer of obscurity, use them sparingly and only with your trusted communication group who understand the designated meanings. Overly complex code words can be difficult to remember and might become confusing during critical situations.

- **Maintaining Physical Security:** Your GMRS radio itself can be a target for theft. Here are some tips to ensure its physical security:
  - o **Secure Storage:** When not in use, store your GMRS radio in a secure location, like a locked backpack or vehicle compartment.
  - o **Password Protection (if available):** If your radio offers password protection functionality, utilize it to deter unauthorized access to programmed channels, privacy codes, or other settings.

Understanding the limitations of GMRS security and implementing these strategies allows you to take control of the level of privacy you desire for your communications. It's important to remember that the goal is to strike a balance between maintaining clear communication within your group and minimizing the risk of sensitive information being intercepted by unintended listeners. This approach not only enhances the security of your GMRS radio communications but also ensures that your communication remains effective and private.

The final section of this chapter will discuss the importance of adhering to regulatory standards and guidelines governing GMRS radio usage. Understanding these regulations ensures responsible operation and avoids potential legal repercussions.

9.5 Compliance with Regulatory Standards and Guidelines: Responsible Operation Within the Rules

Your GMRS radio is a licensed communication tool. To ensure its continued usefulness and avoid encountering legal trouble, it's crucial to understand and comply with the regulations set forth by the Federal Communications Commission (FCC) in the United

States (or the relevant regulatory body in your country). Here's a breakdown of key points to remember:

- **Understanding Your License:** A GMRS license grants you the privilege to operate a GMRS radio within the designated frequency range. Familiarize yourself with the terms and conditions of your license, including limitations on power output, antenna types, and permissible uses.
- **Prohibited Activities:** The FCC prohibits specific activities on GMRS channels, such as:
    - Transmitting false or misleading information.
    - Using obscene or harassing language.
    - Transmitting music, advertisements, or promotional messages.
    - Interfering with authorized communications.
- **Power Output Limitations:** GMRS radios are designed to operate with a specific maximum power output. Exceeding this limit is a violation of FCC regulations. Using unauthorized amplifiers or modifying your radio to boost its power output is strictly prohibited.
- **Maintaining Proper Identification:** While GMRS communication generally doesn't require identifying yourself every time you transmit, it's a good practice to occasionally state your station identification (call sign) to aid in search and rescue efforts or to avoid confusion with other users on the same channel.
- **Staying Updated:** The FCC regulations governing GMRS radio usage can evolve over time. Develop the habit of checking for updates or amendments to the regulations periodically to ensure you remain compliant. The FCC website is a reliable source for current information on GMRS regulations. Simply do a quick

Adhering to these regulatory standards and guidelines demonstrates your commitment to responsible GMRS radio operation. This not only protects you from potential legal consequences but also ensures a more positive and respectful communication environment for all licensed GMRS radio users sharing the channels. By following these guidelines, you contribute to a regulatory framework that promotes safe and lawful communication practices, as outlined in the Federal Communications Commission's regulations. This collective effort not only enhances the overall experience for each user but also fosters a safer and more harmonious communication space, ensuring that GMRS radio remains a reliable tool for staying connected and ensuring safety during your outdoor adventures.

Summary

This chapter transformed you from a basic GMRS radio user into a responsible and informed communicator by exploring essential best practices and safety considerations. Here's a quick recap of the key takeaways:

- **Prioritize Safety:** GMRS is not a substitute for emergencies. Carry a PLB for critical situations and prioritize clear, concise communication during emergencies.
- **Minimize Interference:** Respect shared channels, utilize CTCSS/DCS codes (if available), and be mindful of electronic devices and geographical obstacles that might cause interference.
- **Adapt to Your Environment:** Extreme temperatures, precipitation, altitude, and terrain can impact your radio's performance. Adjust your communication strategies accordingly.
- **Protect Your Privacy (Within Limits):** While GMRS channels aren't inherently secure, consider using CTCSS/DCS codes, minimize sensitive information sharing, and use code words sparingly. Remember, physical security of your radio is important too.
- **Comply with Regulations:** Familiarize yourself with your GMRS license, avoid prohibited activities, adhere to power output limitations, and consider occasionally identifying your station. The FCC website is a valuable resource for current regulations.

Adopting these practices elevates you to the status of a responsible GMRS radio operator, guaranteeing a secure and delightful communication experience for both yourself and others utilizing the shared channels. You are now fully prepared to harness the complete capabilities of your GMRS radio, thereby actively contributing to a harmonious communication atmosphere during your outdoor escapades!

Review Questions

1. Describe two environmental factors that can negatively impact the performance of your GMRS radio and explain strategies you can implement to mitigate these challenges.

2. Imagine you're on a camping trip with a large group spread across several campsites. You need to remind everyone about the importance of respecting the shared GMRS channel and avoiding lengthy conversations. How can you communicate this message effectively while adhering to best practices for clear and concise communication?

3. While GMRS radio communication doesn't require constant identification, explain why it can sometimes be beneficial to occasionally state your station identification (call sign). List two circumstances where stating your call sign might be helpful.

CHAPTER 10

INTEGRATION WITH OTHER RADIO SYSTEMS - EXPANDING THE COMMUNICATION LANDSCAPE

Your GMRS radio serves as a valuable tool for communication in remote locations. However, the world of two-way radio communication extends beyond GMRS. This chapter examines the potential for integrating your GMRS radio with other radio systems, potentially expanding your communication range, functionality, and connecting you with a broader network of users. We'll explore various integration possibilities, but it's important to manage expectations. Some integrations might require additional equipment, technical expertise, or even specific licensing depending on the type of radio system you wish to connect with.

Here's a roadmap of the exciting possibilities we'll explore in this chapter:

- **Compatibility with FRS Radios:** We'll discuss the limitations and potential for communication between GMRS and FRS (Family Radio Service) radios.
- **Connecting GMRS to Repeater Systems:** Learn how GMRS repeaters can extend your communication range and overcome geographical obstacles.
- **Interfacing with Ham Radio Networks:** For the technically-savvy users, we'll explore the complexities of connecting a GMRS radio with amateur (ham) radio networks, potentially granting access to a vast network of communication possibilities.
- **Cross-Band Operation and Interoperability:** This section dives into the concept of using two different radios simultaneously to communicate with users on different frequencies (advanced topic).
- **Networking and Interconnectivity Protocols:** We'll conclude by introducing the concept of networking protocols that facilitate communication between dissimilar radio systems (advanced topic).

By understanding these integration possibilities, you can assess whether expanding your communication horizons beyond the walls of standalone GMRS radio use aligns with your needs and interests. Let's begin on this exploration of potential connections and broaden the landscape of your radio communication experiences!

10.1 Compatibility with FRS Radios: Sharing the Same Frequencies, Different Capabilities

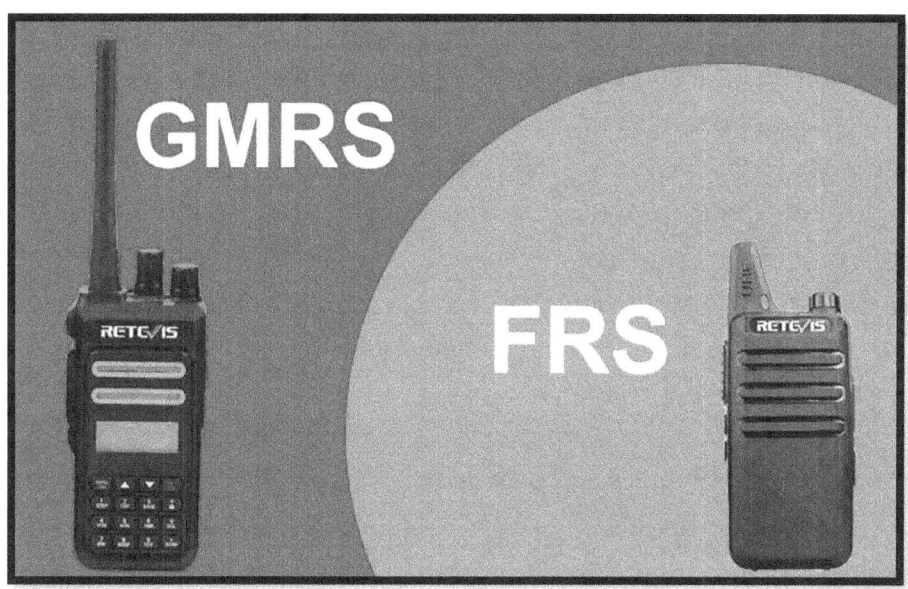

As you learned earlier, GMRS and FRS radios share the same 22 radio frequencies. This might lead you to believe seamless communication is possible between the two. However, there are key distinctions to consider:

- **Power Output:** GMRS radios are generally more powerful than FRS radios, allowing for potentially greater range under ideal conditions. FRS radios are restricted to a maximum power output of 0.5 watts, while GMRS radios can transmit at higher power levels (up to 5 watts for handheld units) depending on the specific model and channel.
- **Privacy Codes (CTCSS/DCS):** While some FRS radios offer CTCSS/DCS functionality, their implementation might not be compatible with GMRS radios due to potential variations in signaling standards. Therefore, relying on CTCSS/DCS codes for privacy between GMRS and FRS radios is not recommended.

- **Channel Limitations:** While both systems share the same frequencies, some FRS radios lack the capability to access all 22 channels. This can limit your ability to communicate with someone using a GMRS radio on a channel not programmed on their FRS radio.

Here's a breakdown of potential communication scenarios:

- **GMRS to FRS:** A GMRS radio can potentially communicate with an FRS radio on a shared channel, provided both radios are tuned to the same frequency and the FRS radio is within range of the GMRS radio's stronger signal. However, the FRS radio user might experience weaker audio quality or reduced range due to the lower power output of their radio.
- **FRS to GMRS:** An FRS radio might be able to receive transmissions from a GMRS radio on a shared channel, but the ability to transmit back and be heard clearly by the GMRS radio is less certain due to the power disparity.

While some level of communication between GMRS and FRS radios might be possible under specific circumstances, it's not always reliable or predictable. For guaranteed clear and consistent communication, using GMRS radios within your group is recommended. If everyone in your communication group upgrades to GMRS radios, you'll benefit from the advantages of increased power output, wider channel availability, and the potential for improved privacy with CTCSS/DCS codes (if your GMRS radios support them).

The next section will explore how GMRS repeaters can extend your communication range and overcome geographical obstacles.

10.2 Connecting GMRS to Repeater Systems: Extending Your Communication Range

While GMRS radios offer a valuable communication tool, their range can be limited by factors like terrain and distance. This section introduces you to GMRS repeaters, powerful tools that can significantly extend your communication reach and overcome geographical barriers.

- **Understanding GMRS Repeaters:**

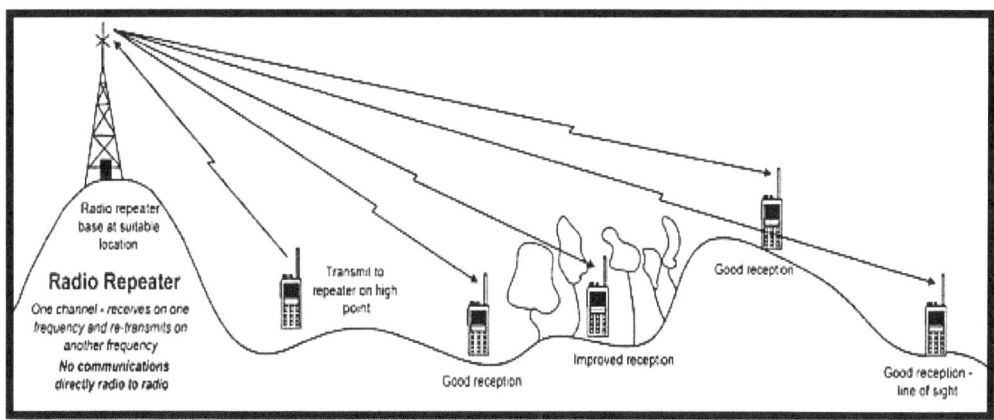

A GMRS repeater is a strategically placed, high-powered radio station that receives transmissions on a specific GMRS channel, amplifies the signal, and retransmits it on another designated frequency. This process effectively extends the range of your GMRS radio communication by relaying signals over a wider area.

- **Benefits of Utilizing Repeaters:**
  - **Extended Range:** Repeaters can significantly increase your communication range, allowing you to connect with others who might be beyond the typical range of your GMRS radio due to geographical obstacles or distance.
  - **Filling in Coverage Gaps:** In areas with challenging terrain like mountains or valleys, repeaters can bridge communication gaps where direct line-of-sight communication might be impossible with a handheld GMRS radio.
  - **Improved Signal Clarity:** Repeaters can amplify weak signals, potentially improving audio quality and clarity, especially at the fringes of your GMRS radio's normal range.
- **Utilizing Repeaters Responsibly:**
  - **Repeater Etiquette:** When using a repeater, adhere to proper etiquette to ensure everyone can benefit from this shared resource. Keep transmissions

concise and avoid lengthy conversations that could clog the channel for others.

- o **Identify Yourself:** It's good practice to identify yourself briefly when using a repeater, stating your call sign or location to aid communication and avoid confusion with other users.
- o **Minimize Background Noise:** Before transmitting, try to minimize background noise around you to ensure your message is clear and easily understandable for others using the repeater.
- **Finding Available Repeaters:** There are several ways to locate GMRS repeaters in your area:
  - o **Online Resources:** Some websites allow you to search for GMRS repeaters by location and view details like their channel frequencies and operating guidelines. Start by searching the internet for these websites, then narrow your search by using the querying tool provided by that specific website.
  - o **Local GMRS User Groups:** Connecting with local GMRS user groups can provide valuable information about repeater availability and proper usage etiquette in your area.
  - o **Consulting Your GMRS Radio Manual:** Some GMRS radio manuals might include pre-programmed repeater frequencies for your region.

Important Considerations:

- **Not all areas have GMRS repeater coverage.** Depending on your location, there might not be readily available GMRS repeaters in your vicinity.
- **Repeaters operate on specific channels.** Ensure your GMRS radio is programmed with the correct repeater input and output frequencies for your desired communication range extension.
- **Licensing Requirements (for some users):** In some countries, operating a GMRS repeater might require additional licensing beyond your standard GMRS license. Always check the regulations in your jurisdiction before setting up or operating a GMRS repeater.

By understanding the benefits and responsible use of GMRS repeaters, you can leverage this technology to significantly expand your communication capabilities during your outdoor adventures. The next section will unravel the complexities of interfacing a GMRS radio with amateur (ham) radio networks.

10.3 Interfacing a GMRS Radio with Ham Radio Networks

The world of amateur (ham) radio offers a vast network of communication possibilities. This section explores the potential for interfacing your GMRS radio with ham radio networks, but it's crucial to understand this is an advanced topic geared towards users with a strong technical background and the appropriate licensing. Here's a breakdown of the complexities involved:

- **Licensing Requirements:** Operating an amateur radio station requires obtaining a ham radio license from the FCC (or the relevant regulatory body in your country). The licensing process involves passing a technical competency exam to demonstrate your understanding of radio theory, operation, and regulations. A GMRS license alone is not sufficient for operating ham radio equipment.

- **Technical Challenges:** Interfacing a GMRS radio with a ham radio network typically involves complex modifications or the use of specialized gateway devices that can translate signals between the different radio systems. This can involve advanced technical knowledge of radio electronics and potentially modifications to your GMRS radio (which could void its warranty).
- **Frequency Restrictions:** GMRS radios operate on a limited set of pre-designated frequencies, while ham radio encompasses a much broader spectrum with various allocations for different license classes. Connecting these two systems requires careful consideration of compatible frequencies and adherence to ham radio regulations regarding allowable communication modes on specific bands.
- **Operational Differences:** Ham radio protocols and communication etiquette differ from those used on GMRS channels. Understanding proper ham radio operating procedures and practices is essential to avoid disrupting communication within the established ham radio networks.

Approaches for Interfacing (For Technically Savvy Users Only):

While not recommended for casual users, here are two potential approaches for interfacing a GMRS radio with ham radio networks, but proceed with caution and only if you possess the necessary technical knowledge and licensing:

- **License a Ham Radio and Utilize a Gateway Device:** Obtaining a ham radio license and utilizing a commercially available gateway device specifically designed to bridge communication between GMRS and ham radio frequencies might be an option. These devices can be expensive and require configuration expertise to ensure proper operation within the legal and technical boundaries of both radio systems.
- **Technical Modifications (Not Recommended):** This approach involves modifying your GMRS radio to operate outside its licensed frequencies, potentially allowing it to transmit or receive on ham radio bands. **This is strongly discouraged** for several reasons:
  - **Legality:** Modifying a radio to operate outside its certified frequencies is a violation of FCC regulations and could result in significant fines or even seizure of your equipment.

- o **Warranty:** Modifying your GMRS radio will likely void its warranty.
- o **Technical Expertise:** Improper modifications can damage your radio or render it inoperable.
- o **Interference:** Operating a modified radio outside its designated frequencies can cause interference with other radio communication systems.

Important Considerations:

- **Interfacing GMRS and ham radio is complex and not for beginners.** Attempting this without proper knowledge and licensing can lead to legal repercussions, damaged equipment, and disruption of communication for other users.
- **Alternative Approaches:** Consider alternative solutions like upgrading your communication group to ham radios (if everyone obtains the necessary licenses) for broader communication possibilities. Ham radio offers a world of features and capabilities beyond the scope of GMRS radios.

Interfacing a GMRS radio with ham radio networks is a complex undertaking best left to experienced and licensed ham radio operators. For most GMRS users, the benefits likely outweigh the risks and complexities involved. The next section will explore the concept of cross-band operation, a technique that utilizes two separate radios for communication on different frequencies (another advanced topic).

10.4 Cross-Band Operation and Interoperability: A Glimpse into Advanced Communication Techniques

The world of radio communication offers a vast array of possibilities beyond the realm of standalone radio use. This section introduces the concept of cross-band operation, an advanced technique that utilizes two separate radios to bridge communication between users operating on different frequencies. While not essential for basic GMRS radio users, understanding this concept can broaden your knowledge of radio communication strategies.

Understanding Cross-Band Operation:

- **Core Principle:** Cross-band operation essentially involves using two radios simultaneously. One radio receives transmissions on a specific frequency (receive frequency), while the other radio transmits on a different frequency (transmit frequency). These two radios can be of the same type (e.g., two GMRS radios) or completely different radio systems (e.g., a GMRS radio and a ham radio).
- **Applications:** Cross-band operation can be useful in various scenarios:
  - **Extending Communication Range:** By utilizing a powerful base station radio for transmitting on a different frequency, you can potentially extend the communication range of a lower-powered handheld radio receiving on another frequency. This can be helpful in situations where a direct line-of-sight connection might be challenging or where one radio has a limited range.
  - **Relaying Signals:** Cross-band operation can be used to relay signals between two locations that cannot communicate directly due to geographical obstacles or incompatible radio systems. For example, a GMRS radio receiving on a local channel can be used to relay messages to a ham radio transmitter on a different frequency that can reach a distant location.
- **Technical Considerations:** Cross-band operation can involve complexities such as:
  - **Dual Radio Setup:** You'll need two separate radios, each programmed for their respective transmit and receive frequencies.
  - **Synchronization:** Ensuring both radios operate in a coordinated manner to facilitate seamless message relay can require technical expertise.
  - **Frequency Legality:** It's crucial to ensure you're using frequencies within your licensed radio privileges and adhering to regulations for each radio system involved in the cross-band operation.
  - **Operational Complexity:** Mastering the coordination and timing involved in effective cross-band communication requires practice and a strong understanding of radio protocols.

Interoperability: A Related Concept:

Closely related to cross-band operation is the concept of interoperability. Interoperability refers to the ability of different radio systems to communicate with each other. While true interoperability often involves complex protocols and specialized equipment, cross-band operation can sometimes be utilized as a rudimentary form of interoperability by establishing a communication bridge between two dissimilar radio systems.

Important Considerations:

- **Cross-band operation is an advanced technique.** For most GMRS users, focusing on mastering clear and efficient communication within the GMRS system itself is the most practical approach.
- **Licensing Requirements:** Depending on the specific frequencies used in your cross-band operation setup, you might require additional licenses beyond your basic GMRS license. Always check the regulations in your jurisdiction.

Conclusion:

Cross-band operation offers a glimpse into the intricate world of advanced radio communication techniques. While not essential for everyday GMRS use, understanding this concept can spark your curiosity and inspire you to explore the vast potential of radio communication systems. The next section will conclude this chapter by introducing networking protocols, another advanced topic that facilitates communication between dissimilar radio systems.

10.5 Networking and Interconnectivity Protocols: The Language of Complex Radio Systems

This final section of Chapter 10 ventures into the realm of networking protocols, a complex topic that governs communication between dissimilar radio systems. While understanding the intricacies of networking protocols might not be necessary for basic GMRS radio use, it provides valuable insight into the underlying mechanisms that enable communication across different radio technologies.

- **Networks and Communication Protocols:** A network can be thought of as a collection of interconnected devices that can share resources and information. In the context of radio communication, networks can encompass various radio systems that need to exchange data. Networking protocols define a set of rules and specifications that govern how these radio systems communicate with each other.
- **Examples of Networking Protocols in Radio Communication:**
  - **Digital Mobile Radio (DMR):** A popular digital radio system used for commercial and professional applications. DMR utilizes specific protocols to ensure data packets are formatted and transmitted correctly between DMR radios on the network.
  - **Packet Radio:** A communication method that transmits data in small packets rather than continuous streams. Packet radio protocols define how these data packets are addressed, routed, and error-corrected to ensure reliable communication across radio networks.
- **The Role of Networking Protocols in GMRS Systems (Limited Application):**
  - **Minimal Networking in Basic GMRS Operation:** Standalone GMRS radio communication typically doesn't involve complex networking protocols. GMRS radios operate on pre-designated channels and rely on relatively simple protocols to establish communication within the GMRS system itself.
  - **Potential Future Applications:** As technology evolves, future GMRS radios might incorporate networking features that utilize protocols to enable more advanced functionalities like interconnectivity with other radio systems or data transfer capabilities.
- **Importance of Understanding Protocols (for Developers):**
  - **Advanced Applications:** For developers creating applications or interfaces that interact with GMRS radios or other radio systems, understanding the underlying networking protocols is crucial for ensuring compatibility and proper data exchange.

The concept of networking protocols lays the foundation for complex communication systems that enable data exchange across diverse radio technologies. While not essential for everyday GMRS radio use, this knowledge can equip developers and innovators with the tools to explore the future possibilities of GMRS integration with broader communication networks.

Do not forget, the level of complexity involved in these integrations can vary greatly. For most users, focusing on mastering clear and efficient communication within the GMRS system itself is a solid foundation. However, the knowledge you've gained empowers you to explore the exciting potential of GMRS radio and its potential connection to the broader world of radio communication.

Summary

This chapter explored the potential for expanding your GMRS radio's usefulness by integrating it with other radio systems. Here's a quick recap:

- **Limited Compatibility with FRS Radios:** While FRS and GMRS radios share frequencies, communication can be unreliable due to power output differences and potential channel limitations on some FRS radios.
- **Extending Range with GMRS Repeaters:** GMRS repeaters can significantly boost your communication range by amplifying and retransmitting signals over a wider area. Learn proper repeater etiquette and locate repeaters in your area for extended reach.
- **Interfacing with Ham Networks (Advanced Topic):** Connecting a GMRS radio to ham radio networks requires a ham radio license, technical expertise, and adherence to complex regulations. For most users, the challenges outweigh the benefits.
- **Cross-Band Operation (Advanced Topic):** This technique utilizes two separate radios to communicate on different frequencies. It can extend range or relay signals but involves technical complexities and potential licensing requirements.
- **Networking Protocols (Advanced Topic):** These protocols govern communication between dissimilar radio systems. While not crucial for basic GMRS use, they form the foundation for complex radio networks.

Understanding these integration possibilities allows you to assess if expanding your communication horizons aligns with your needs. For most users, mastering clear and efficient GMRS communication is the primary focus. The knowledge gained here empowers you to explore the potential for connecting your GMRS radio to a broader world of radio communication possibilities.

Review Questions

1. Imagine you're on a hiking trip and encounter another group using FRS radios. They're outside of the range of your GMRS radio's direct communication, but you see a GMRS repeater tower in the distance. Explain how you could potentially communicate with the FRS group using your GMRS radio and the repeater.

2. You're interested in the concept of extending your communication range beyond the limitations of a standalone GMRS radio. Briefly describe two potential approaches you could explore, and explain the key considerations or limitations associated with each approach.

3. Networking protocols are a complex topic, but understanding the basic concept can be beneficial. In your own words, explain the main purpose of networking protocols in the context of radio communication systems.

CHAPTER 11

UNVEILING THE VERSATILITY OF GMRS RADIOS - APPLICATIONS AND CASE STUDIES

We've discussed the technical aspects of GMRS radios, exploring their functionalities, operational considerations, and potential for integration with other radio systems. Now, it's time to explore the many ways GMRS radios can be applied in real-world scenarios. This chapter will showcase the versatility of GMRS radios through a variety of use cases, from personal and family communication to business applications, emergency response, and even recreational activities. By exploring these diverse applications and real-life case studies, you'll gain a deeper understanding of the value and potential impact that GMRS radios can have in various aspects of our lives.

Here's a roadmap of the exciting applications we'll explore in this chapter:

- **Personal and Family Use Cases:** Discover how GMRS radios can enhance communication and safety within your family group during everyday activities or emergencies.
- **Business and Commercial Applications:** Learn how businesses can leverage GMRS radios for improved coordination, logistics, and safety in various commercial settings.
- **Public Safety and Emergency Services:** Explore the role of GMRS radios as a valuable tool for first responders and emergency management personnel.
- **Recreational and Outdoor Adventures:** See how GMRS radios can elevate your outdoor experiences by providing a reliable communication channel for safety, coordination, and enjoyment.
- **Military and Government Applications:** While this section will provide a brief overview, due to the specialized nature of these applications, the focus will remain on civilian GMRS radio uses.

By the end of this chapter, you'll have a comprehensive understanding of the diverse applications for GMRS radios and how they can be a valuable asset in various situations. So, let's explore the true potential of these versatile communication tools!

11.1 Personal and Family Use Cases: Strengthening Communication and Safety for Your Loved Ones

GMRS radios shine as communication tools for families and individuals seeking reliable and efficient ways to stay connected. Here are some key personal and family use cases that highlight the valuable role GMRS radios can play:

- **Enhancing Coordination During Outings and Adventures:** Whether you're enjoying a day at the park, exploring a hiking trail, or on a family camping trip, GMRS radios provide a dedicated communication channel to stay connected with your family members. This can be crucial for coordinating activities, ensuring everyone's safety, and maintaining peace of mind, especially for larger groups spread out across an area.

- **Maintaining Contact During Emergencies:** In unforeseen circumstances like emergencies or getting separated from your group, a GMRS radio can be a lifeline. The ability to communicate your situation and location to other family members with GMRS radios can be critical for a prompt response and ensure everyone's safety during emergencies.

- **Improving Communication at Home and in Large Properties:** For families living on large properties or with sprawling backyards, GMRS radios can be a convenient way to stay connected and coordinate tasks or chores around the house. They can also be helpful for quick communication between family members without needing to rely on cell phones, which might not always have reliable reception in certain areas.

- **Peace of Mind for Parents with Active Children:** For parents with children who play outdoors or participate in activities where direct supervision might not always be possible, GMRS radios can offer a sense of security. Knowing you can easily communicate with your children through a dedicated channel can provide peace of mind and allow for quick check-ins or emergency contact.

- **Safety During Solo Activities:** Even for solo adventurers venturing outdoors, a GMRS radio can be a valuable safety tool. The ability to call for help in case of an emergency or unexpected situation can provide a sense of security and potentially lead to a faster response time from emergency services or nearby individuals with GMRS radios.

Important Considerations for Personal and Family Use:

- **Understanding GMRS Regulations:** Familiarize yourself with the limitations and responsible use guidelines outlined in your GMRS license. Remember, GMRS channels are shared, so courtesy and clear communication are essential.

- **Developing a Communication Plan:** Establish a basic communication plan with your family members regarding channel usage, emergency protocols, and appropriate call signs or identifiers for clear communication within your family group.

- **Practice Makes Perfect:** Don't wait for an emergency to become familiar with your GMRS radios. Practice using them regularly with your family to ensure everyone understands proper operation and feels comfortable communicating through this channel.

Integrating GMRS radios into your family's communication strategy not only enhances coordination but also significantly improves safety during outings and emergencies. This integration fosters a sense of connection and security for all your loved ones, ensuring that you are well-prepared for any situation. In the following section, we will discuss how businesses can harness the capabilities of GMRS radios to significantly improve their communication and coordination efforts.

11.2 Business and Commercial Applications: Boosting Efficiency and Safety in the Workplace

Beyond personal and family use, GMRS radios offer a range of valuable applications for businesses and commercial operations. Here's a closer look at how GMRS radios can enhance communication, coordination, and safety in various commercial settings:

- **Improved Communication on Worksites:** In construction zones, event venues, or other large outdoor work areas, GMRS radios can provide a reliable and instant communication channel for workers spread across the site. This facilitates efficient task coordination, faster response times to issues, and improved overall teamwork.
- **Enhanced Logistics and Delivery Operations:** Delivery companies and transportation services can benefit from GMRS radios for real-time communication

between drivers and dispatchers. This allows for route updates, delivery confirmations, and improved coordination for a more streamlined logistics operation.

- **Streamlined Security and Patrol Activities:** Security personnel patrolling warehouses, properties, or event venues can utilize GMRS radios for quick communication and coordinated response to security incidents. The ability to report suspicious activity, request backup, or share updates on patrol routes can significantly enhance security effectiveness.
- **Communication During Public Events:** Event organizers responsible for large gatherings like concerts, festivals, or sporting events can leverage GMRS radios for seamless communication between staff members. Coordinating logistics, managing crowd control, and ensuring everyone is on the same page becomes more efficient with a dedicated communication channel.
- **Improved Customer Service in Hospitality:** Resorts, hotels, or other hospitality businesses can utilize GMRS radios for staff communication to enhance customer service. Coordinating housekeeping requests, expediting guest assistance, or providing real-time communication between departments can lead to a more efficient and responsive guest experience.

Important Considerations for Business Use:

- **Licensing Requirements:** Businesses utilizing GMRS radios need to ensure they comply with licensing regulations. A single GMRS license might not suffice for extensive business use. Exploring the possibility of obtaining a Part 90 business radio license from the FCC might be necessary depending on the scale and complexity of your communication needs.
- **Channel Etiquette and Professionalism:** Maintain professional communication protocols on GMRS channels shared with personal users. Be mindful of channel congestion, keep transmissions concise and clear, and avoid using profanity or inappropriate language.
- **Integration with Other Systems:** For complex business communication needs, explore the potential for integrating GMRS radios with other communication systems like phone lines or dispatch software for a more comprehensive communication network.

Incorporating GMRS radios into their communication infrastructure allows businesses to significantly enhance operational efficiency, improve coordination among employees, and ultimately deliver superior service to their clients or customers. The following section will explore the pivotal role of GMRS radios in public safety and emergency response situations, underscoring their invaluable contribution to effective communication and coordination during critical times.

11.3 Public Safety and Emergency Services: A Reliable Communication Tool in Critical Situations

When emergencies strike or public safety is at stake, reliable and efficient communication becomes paramount. GMRS radios, while not a primary communication tool for all first responders, can serve as a valuable asset in specific situations for various public safety agencies and emergency response personnel. Here's a detailed exploration of how GMRS radios can play a crucial role in public safety:

- **Supporting Search and Rescue Operations:** Search and rescue teams operating in remote locations or areas with limited cellular coverage can leverage GMRS radios for on-site communication. Coordinating search efforts, reporting updates on missing persons, and maintaining clear communication between team members can be critical for a successful search and rescue operation.

- **Incident Management and Crowd Control:** During emergencies or large public gatherings where maintaining order and managing crowds is essential, GMRS radios can provide a dedicated channel for event staff, security personnel, and first responders to coordinate their response. This can be crucial for ensuring efficient crowd control, directing emergency personnel to critical areas, and facilitating a swift response to evolving situations.
- **Communication During Natural Disasters:** In the aftermath of natural disasters that might disrupt traditional communication infrastructure like cell phone towers, GMRS radios can offer a backup communication channel for emergency responders. This allows for relaying critical information, coordinating relief efforts, and maintaining communication between teams working in disaster zones.
- **Auxiliary Communication for Fire Departments:** While fire departments primarily rely on dedicated communication systems, GMRS radios can sometimes serve as a supplementary communication tool. For instance, they might be used by firefighters on the scene of a large fire to maintain communication with incident commanders or other teams operating in different areas.
- **Community Emergency Response Teams (CERTs):** Civilian volunteers trained in basic emergency response procedures can utilize GMRS radios to assist professional first responders during emergencies. This can be helpful for relaying information, supporting crowd control efforts, or assisting with communication needs within the community.

Important Considerations for Public Safety Use:

- **Interference with Primary Communication Systems:** It's crucial to understand that GMRS radios operate on shared frequencies. In situations where emergency responders utilize dedicated communication systems, GMRS radio use should be limited to avoid potential interference.
- **Limited Range:** While GMRS radios offer improved range compared to some personal communication devices, their range can still be limited by terrain or distance. Public safety professionals should be aware of these limitations and rely on primary communication systems when necessary.
- **Licensing and Training:** For public safety personnel using GMRS radios, proper licensing and training are essential. Understanding GMRS regulations, proper

communication protocols, and potential limitations is crucial for ensuring effective and responsible use in emergency situations.

Complementary Role to Primary Systems:

It's important to remember that GMRS radios are not a replacement for dedicated emergency response communication systems used by fire departments, police forces, or other primary responders. However, in specific scenarios and as a supplementary communication tool, GMRS radios can play a valuable role in supporting public safety efforts and facilitating communication during critical situations.

The next section will explore the exciting world of GMRS radios in the context of recreational activities and outdoor adventures.

11.4 Recreational and Outdoor Adventures: Elevating Your Experience with Reliable Communication

The great outdoors beckon, offering opportunities for exploration, adventure, and connecting with nature. GMRS radios can become an invaluable companion on your recreational pursuits, enhancing your experience in several ways:

- **Safety on Hiking Trails and Backpacking Trips:** Whether venturing on a solo hike or a group backpacking adventure, GMRS radios provide a reliable communication channel, especially in remote areas with limited cell phone service. The ability to call for help in case of emergencies like injuries, getting lost, or encountering unexpected situations can provide peace of mind and potentially lead to a faster response time from search and rescue teams or nearby individuals with GMRS radios.

- **Improved Group Coordination During Camping Trips:** For larger camping groups spread across a campsite, GMRS radios offer a dedicated channel for coordinating activities, ensuring everyone's safety, and staying connected. This can be helpful for tasks like assigning camp chores, keeping track of group members exploring the surroundings, or simply staying in touch for added enjoyment and peace of mind.

- **Communication on Off-Road Adventures:** For off-road enthusiasts navigating trails in vehicles like Jeeps or ATVs, GMRS radios offer a reliable communication tool for maintaining contact with fellow adventurers. Coordinating routes, reporting obstacles or hazards on the trail, and ensuring everyone stays together becomes easier with a dedicated communication channel, especially in areas with limited cell phone reception.

- **Enhanced Communication During Boating and Fishing Activities:** While boaters and fishers typically rely on VHF marine radios for primary communication on designated channels, GMRS radios can offer a supplementary communication tool for specific situations. For instance, they can be used for short-range communication between boats in your group or for maintaining contact with onshore personnel while exploring waterways. **Remember, GMRS radios should not be used as a primary communication tool for emergencies at sea. Always rely on VHF marine radios for critical communication needs while boating.**

- **Search and Rescue Operations (Civilian Role):** If you're involved in organized search and rescue efforts coordinated with trained professionals, GMRS radios might be used as a supplementary communication tool for civilian volunteers. Following proper training and adhering to guidelines set by the search and rescue team leader, GMRS radios can facilitate communication between volunteers and support search efforts in collaboration with primary responders.

Important Considerations for Recreational Use:

- **Understanding GMRS Regulations:** Familiarize yourself with the limitations and responsible use guidelines outlined in your GMRS license. Be mindful of channel congestion, especially in popular outdoor areas, and keep transmissions concise and clear to avoid disrupting communication for others.
- **Respecting Channel Etiquette:** Maintain a courteous and professional tone on shared GMRS channels. Avoid excessive chatter or using inappropriate language, especially in areas where other recreational users might be relying on the same channels.
- **Knowing Your Limitations:** GMRS radios offer improved range compared to cell phones, but their range can still be limited by terrain or distance. Be aware of these limitations and plan your communication strategies accordingly. Carry additional safety devices like personal locator beacons (PLBs) for emergency situations in remote areas.

By incorporating GMRS radios into your recreational activities, you can enhance safety, improve group coordination, and add a layer of security to your outdoor adventures. The next section will provide a brief overview of GMRS radio applications in military and government contexts.

11.5 Military and Government Applications: A Focus on Civilian Use

The world of military and government communication systems is vast and complex, utilizing a variety of specialized radio technologies beyond the scope of civilian GMRS radios. However, this section provides a brief glimpse into potential limited applications of GMRS radios in these sectors, while emphasizing the importance of focusing on civilian use cases throughout this chapter.

Limited Military and Government Use:

- **Military Maneuvers (Simulations):** During non-critical military training exercises, GMRS radios might be used for short-range communication between simulated patrols or teams within a limited training area. This allows for practicing basic communication protocols and coordination techniques in a controlled environment without relying on dedicated military communication systems.

- **Supplemental Communication for Non-Critical Tasks:** In some instances, government agencies or military personnel might utilize GMRS radios for non-critical communication tasks on a limited basis. For example, park rangers coordinating with volunteers on maintenance projects in remote areas might find GMRS radios a helpful tool.

Important Considerations:

- **Security Concerns:** Military and government communication needs often involve sensitive information. GMRS radio communication is not encrypted, making it unsuitable for transmitting classified information.
- **Limited Range and Capabilities:** The range and feature set of GMRS radios are not comparable to dedicated military communication systems designed for secure and long-range communication in complex operational environments.
- **Focus on Civilian Use:** The primary focus of this chapter is on civilian applications of GMRS radios. While briefly mentioned here for the sake of completeness, military and government applications are far beyond the scope of this exploration.

Civilian Use in Support of Government Efforts:

- **Community Emergency Response Teams (CERTs):** As mentioned earlier, civilian volunteers participating in CERT programs coordinated with local authorities might utilize GMRS radios as a supplementary communication tool during emergencies. This can be helpful for relaying information or supporting public safety efforts under the guidance of trained professionals.
- **Search and Rescue (Civilian Role):** Similar to the concept mentioned in the recreational use section, civilian volunteers involved in organized search and rescue operations might use GMRS radios for communication purposes, following proper training and adhering to protocols set by the search and rescue team leader.

While GMRS radios might have very limited applications in specific military and government contexts, their primary value lies in enhancing communication for civilians in various personal, professional, and recreational scenarios. By understanding the diverse applications explored throughout this chapter, you can leverage the potential of GMRS radios to improve safety, coordination, and overall communication effectiveness in a variety of situations.

CHAPTER 11 SUMMARY

UNVEILING THE POWER OF GMRS RADIOS - A MULTIFACETED TOOL FOR COMMUNICATION

This chapter explored the diverse applications of GMRS radios, showcasing their potential to enhance communication and safety in various aspects of our lives. Here's a quick recap of the key areas covered:

- **Personal and Family Use Cases:** GMRS radios provide a reliable communication channel for families, improving coordination during outings, maintaining contact in emergencies, and fostering a sense of security, especially for those with active children or large properties.

- **Business and Commercial Applications:** Businesses can leverage GMRS radios to streamline communication on worksites, enhance logistics and delivery operations, improve security patrol effectiveness, ensure smooth communication during events, and provide better customer service in hospitality settings.

- **Public Safety and Emergency Services:** While not a primary communication tool, GMRS radios can serve as a valuable backup in specific situations. They can support search and rescue operations, facilitate communication during crowd control or natural disasters, and act as a supplementary tool for fire departments or CERT volunteers.

- **Recreational and Outdoor Adventures:** For outdoor enthusiasts, GMRS radios enhance safety on hikes or camping trips, improve group coordination, facilitate communication during off-road adventures, and provide a supplementary communication channel for boaters and fishers (remembering VHF marine radios are primary for emergencies at sea).

- **Military and Government Applications (Limited Focus):** While military communication is complex and utilizes specialized systems, GMRS radios might be used for limited purposes in training simulations or non-critical government communication tasks. The emphasis throughout this chapter remains on civilian applications of GMRS radios.

By understanding these diverse applications and adhering to responsible use guidelines, you can unlock the true potential of GMRS radios and transform them into a valuable communication asset for various personal, professional, and recreational pursuits.

Review Questions

1. Imagine you work for a park ranger service and are coordinating a team of volunteers on a trail maintenance project in a remote area. Cell phone service is unreliable in this location. Discuss the pros and cons of utilizing GMRS radios for communication between volunteers and park rangers during this project.

2. You're a backpacker planning a solo trip in a national park. You're torn between bringing a GMRS radio for potential emergencies or opting for a lighter weight setup without the radio. Explain the factors you would consider when making this decision, referencing the potential benefits and limitations of GMRS radios in a wilderness setting.

3. You and your friends are planning a kayaking trip on a local lake. While you all have VHF marine radios for primary communication on designated channels, you're also considering bringing GMRS radios for short-range communication between your kayaks. Explain the potential benefits and drawbacks of using GMRS radios in this scenario, emphasizing the importance of adhering to safety regulations.

Conclusion: The Empowering World of GMRS Radios - A Journey of Communication and Connection

Congratulations! You've reached the conclusion of this comprehensive exploration of GMRS radios. Throughout this book, we've explored the technical aspects of these radios, unpacked their functionalities, explored regulations and licensing requirements, and most importantly, unveiled the vast potential applications of GMRS radios in various facets of our lives.

From enhancing communication and safety for families to streamlining operations in businesses, from supporting public safety efforts to elevating the enjoyment of outdoor adventures, GMRS radios offer a versatile communication tool with a wide range of use cases. By understanding their capabilities and limitations, you can make informed decisions about incorporating GMRS radios into your communication strategy.

Remember, responsible use is paramount. Familiarize yourself with GMRS regulations, adhere to proper channel etiquette, and prioritize safety in all your communication endeavors. As you venture out with your GMRS radio, you not only enhance your ability to connect with others but also contribute to a more informed and coordinated community, fostering a shared sense of security and responsibility.

This book has equipped you with the knowledge to leverage the power of GMRS radios. Whether you use them for personal communication, business operations, or outdoor activities, remember, GMRS radios are more than just technological tools. They are gateways to improved communication, stronger connections, and a heightened sense of security for yourself and those around you.

So, the next time you embark on an adventure, head out to a worksite, or simply want to stay connected with loved ones, consider the potential of GMRS radios. With responsible use and an understanding of their capabilities, you can unlock a world of communication possibilities and transform your GMRS radio into a valuable asset for navigating the ever-evolving landscape of connection in our modern world.

INDEX

www.ingramcontent.com/pod-product-compliance
Lightning Source LLC
Chambersburg PA
CBHW082106220526
45472CB00009B/2072